Literary Urban Studies

The Literary Urban Studies Series has a thematic focus on literary mediations and representations of urban conditions. Its specific interest is in developing interdisciplinary methodological approaches to the study of literary cities. Echoing the Russian formalist interest in literaturnost or literariness, Literary Urban Studies will emphasize the "citiness" of its study object—the elements that are specific to the city and the urban condition—and an awareness of what this brings to the source material and what it implies in terms of methodological avenues of inquiry. The series' focus allows for the inclusion of perspectives from related fields such as urban history, urban planning, and cultural geography. The series sets no restrictions on period, genre, medium, language, or region of the source material. Interdisciplinary in approach and global in range, the series actively commissions and solicits works that can speak to an international and cross-disciplinary audience.

More information about this series at
http://www.palgrave.com/gp/series/15888

Maria Beville • Deirdre Flynn
Editors

Irish Urban Fictions

palgrave
macmillan

Editors
Maria Beville
Centre for Studies in Otherness
Aarhus University, Denmark

Deirdre Flynn
University College Dublin
Dublin, Ireland

ISSN 2523-7888 ISSN 2523-7896 (electronic)
Literary Urban Studies
ISBN 978-3-319-98321-9 ISBN 978-3-319-98322-6 (eBook)
https://doi.org/10.1007/978-3-319-98322-6

Library of Congress Control Number: 2018951602

Cover illustration: The quays at twilight in Dublin City, Ireland
Credit: David Soanes Photography / Getty Images

This Palgrave Macmillan imprint is published by the registered company Springer Nature Switzerland AG
The registered company address is: Gewerbestrasse 11, 6330 Cham, Switzerland

Acknowledgements

The editors would like to thank the contributors to this volume for their intellectual rigour, cordiality, and professionalism throughout the entire process. Overall it has been an exciting project and has inspired new research projects for many of us. It has been a pleasure working with such an excellent team of scholars.

We would also like to thank those who were involved in peer reviewing the work and offering invaluable insights at various stages from abstract to completed manuscript.

Deirdre Flynn would like to thank her colleagues in the University College Dublin (UCD) School of English, Drama, Creative Writing and Film for their support and friendship.

Maria Beville would like to thank the team at the Centre for Studies in Otherness for encouraging new ideas and promoting ongoing work on literature and the urban.

Contents

Notes on Contributors

Maria Beville is a researcher, lecturer, and writer with the Centre for Studies in Otherness. Her research interests include the Gothic, Irish Studies, and urban literary studies. Working mostly with contemporary fiction and film, her recent research has focused on the supernatural city in literature. Her books include *The Unnameable Monster in Literature and Film* (2013), *The Gothic and the Everyday* (coedited; 2014), and *Gothic-postmodernism* (2009). She is editor of the journal *Otherness: Essays and Studies.*

Martyn Colebrook completed his PhD in 2012 focusing on the novels of Iain Banks. To date he has delivered over 100 conference papers in the UK, Ireland, the USA, and Europe. He has published chapters focusing on Don DeLillo, China Miéville, and Gordon Burn and contributed to the Bloomsbury Decades Series on Contemporary Literature. He co-edited the first collection of scholarly essays on Ian Banks' fiction and has co-organised conferences on Michael Moorcock, 9/11 Narratives, Millennial Fictions, Angela Carter, and Jeanette Winterson.

Molly Ferguson is an Assistant Professor of English at Ball State University in Indiana, where she teaches courses in Postcolonial literatures and Women's and Gender Studies. Her primary research interest is in contemporary Irish literature and its intersections with gender studies, human rights, austerity, and trauma theory. She has published articles in *Journal of Postcolonial Writing*, *Women's Studies: An Interdisciplinary Journal*, *New Hibernia Review*, *Studi Irlandesi*, *The Canadian Journal of Irish Studies*, and *Nordic Irish Studies.*

Deirdre Flynn is Assistant Professor of Irish Studies in the School of English, Drama, Creative Writing and Film at University College Dublin. Her research interests are in Contemporary and World Literature, Irish Studies, Drama and Theatre, Urban Studies, Dystopian fiction, ageing, and gender. She recently worked on a co-edited collection titled *Representations of Loss in Irish Literature* (Palgrave Macmillan, 2018) and is currently working on a monograph on Haruki Murakami.

Nikhil Gupta is an Academic Standards Board Advisor at the University of Michigan. His research looks at stories of empires that Irish and American modernist writers rewrite and transpose from one side of the Atlantic to the other in order to reimagine belonging in their own homelands. This transnational framework casts revision and its crucial role in the modernist practice of artistic production as an intertextual process, one which also places British imperialism alongside American expansion and views them as parts of the same geopolitical process.

Tim Keane is an Associate Professor of English at Borough of Manhattan Community College in the City University of New York. He specialises in the intersection between the visual arts and literature in twentieth-century modernism, both European and American. He has published articles on the writings of artist Ray Johnson and the composer and musician John Cage, on painter George Schneeman's collaborations with the 'New York School' poets, and on painter and poet Joe Brainard's memoirs.

Laura Lovejoy is a postdoctoral fellow in the School of English at University College Cork. With funding from the Irish Research Council, she is preparing a monograph, *Modernism's Red Lights: London, New York, Dublin, Berlin*, which explores modernist geographies of commercial sex. Her work has been published in *Humanities* and the *Journal of Working-Class Studies*.

Neil Murphy is an Associate Professor of English at Nanyang Technological University (NTU), Singapore. He is the author of *Irish Fiction and Postmodern Doubt* (2004) and editor of *Aidan Higgins: The Fragility of Form* (2010). He co-edited (with Keith Hopper) a special Flann O'Brien centenary issue of the *Review of Contemporary Fiction* (2011) and *The Short Fiction of Flann O'Brien* (2013) and a four-book series related to the work of Dermot Healy, all with Dalkey Archive Press, USA, including, most recently, *Writing the Sky: Observations and Essays* *on Dermot Healy* (2016). His book on John Banville is published by Bucknell University Press in 2018.

Quyen Nguyen has recently completed her PhD with the Division of English at NTU, Singapore. Her research interests include James Joyce, Irish literature, modernism, translation studies, and geocriticism. Her PhD dissertation is entitled 'City as Writing: Textual Dublin in *Ulysses*'.

Margaret O'Neill is a Gender Arc Project Coordinator at the University of Limerick. In summer 2018, she visited NUI Galway as a Moore Institute visiting fellow to develop a project on women, ageing, and life narratives. Margaret has taught Irish literature at the University of Limerick and Maynooth University. She researches in twentieth-century and contemporary Irish writing, gender theory, health humanities, and cultural gerontology. Her co-edited collection *Ageing Women in Literature and Visual Culture: Reflections, Refractions, Reimaginings* was published by Palgrave in 2017.

Terry Phillips was the Dean of Arts and Humanities at Liverpool Hope University until her retirement in 2010 and continues to chair the Irish Studies Research Group. Her recent publications include *Irish Literature and the First World War* (Peter Lang, 2015); 'Out on a Great Adventure: The Travels of Patrick MacGill' in *Travel in France and Ireland: Tourism, Sport and Culture*, eds Francis Healy and Brigitte Bastiat (Peter Lang, 2017); and 'Our Dead Shall Not Have Died in Vain: The War Poetry of Harry Midgley' in *Towards 2016: 1916 in Irish Literature, Culture and Society*', eds Crosson and Huber (EFACIS, 2015).

Dawn Miranda Sherratt-Bado is an academic and a dual specialist in Irish and Caribbean Studies. She is a co-editor of *Female Lines: New Writing by Women from Northern Ireland* (New Island Books, 2017). She is the author of *Decoloniality and Gender in Jamaica Kincaid and Gisèle Pineau: Connective Caribbean Readings* (Palgrave Macmillan, 2017). Dawn has also published in *Irish Studies Review*, *Breac*, *Callaloo*, *The Sunday Business Post*, *Four Nations History*, and *Writing the Troubles*. She is a regular contributor to *The Honest Ulsterman*, the *Dublin Review of Books*, and *The Irish Times*.

Eva Roa White is a Professor of English at Indiana University Kokomo, USA. Her research focuses primarily on Irish and immigration studies. She has authored a book, *A Case Study of Ireland and Galicia's Parallel Paths* *to Nationhood*. Her articles about identity migration and hybrid cultural and national identities in Irish, Galician, and South Asian Studies have appeared in *New Hibernia Review* and *South Asian Review* and in the essay collections *Literary Visions of Multicultural Ireland: The Immigrant in Contemporary Irish Literature*, *The Wake of the Tiger*, and *(M)Othering the Nation: Constructing and Resisting Regional and National Allegories Through the Maternal Body*.

CHAPTER 1

Introduction: Irish Urban Fictions

Maria Beville and Deirdre Flynn

'You take delight not in a city's seven or seventy wonders, but in the answer it gives to a question of yours … Or the question it asks you, forcing you to answer, like Thebes through the mouth of the Sphinx'
(Italo Calvino—Le città invisibili)

The first UNESCO 'Cities of Literature Conference' was hosted in Dublin in June 2016 presenting the Irish capital as a creative, literary city and celebrating Ireland's literary icons. Claiming that Dublin 'has words in its blood', the project explored Ireland's literary traditions across a range of urban settings and continued work on the importance of literature to Dublin which has been ongoing since the city was designated as a UNESCO City of Literature in 2010.[1] Literature is of great significance not only to Dublin but to all Irish cities. Each of Ireland's ten cities, both north and south of the border, has its own recognisable literary heritage which has evolved over time and through a variety of authors and literary

M. Beville (✉)
Centre for Studies in Otherness, Aarhus University, Denmark

D. Flynn (✉)
University College Dublin, Dublin, Ireland

M. Beville, D. Flynn (eds.), *Irish Urban Fictions*, Literary Urban Studies, https://doi.org/10.1007/978-3-319-98322-6_1

styles. Equally, the cities themselves have been and are important to Irish literature and have inspired some of the country's finest writing.

Joseph Valente has noted that 'the urban experience has indeed been comparatively underappreciated in studies of Irish literature'.[2] Contemporary literary scholars such as Gerry Smyth have offered nuanced and important readings of Dublin's urban literary contexts and the wider cultural significance of these. However, a broader examination of the range of Irish literary cities has yet to be completed. By virtue of their location on a small island on the periphery of Western Europe, Irish cities offer a unique urban cultural experience distinct from that of frequently fictionalised megacities such as London, Paris, and Tokyo. Understanding the cultural, social, and political mosaic that comprises Irish cities, north and south of the border, and the writing that they inspire, can open up new perspectives, not just on Irish literature, but on the broader discourse of 'the global city' and the city as a response to capitalist advancement and economy in flux.[3]

Reflected in recent, but also in late nineteenth- and early twentieth-century Irish fictions, the topography of the Irish city[4] is varied and complex. According to Smyth, Joyce's Dublin was a composite of the various layers which had contributed to the emergence of the modern city up to the late nineteenth century—Viking trading town, colonial buttress, Georgian capital, industrial slum, and was also, in the words of Declan Kiberd 'a classic example of a periphery dominated centre'. Since then, Smyth argues, the city has continued to exist as 'a complex living entity moving in time as well as in space'.[5] Elaborating further, Smyth argues that Dublin is not just a combination of physical and imagined environments, it is a 'word city',[6] and the words that create it and emerge from it engender an urban narrative as fluid as the materiality of the city itself. Building on Smyth's approach to the constantly evolving 'urban fabric' of the city, we extend analysis of the living city in Irish literature to include cities outside of Dublin which have not been subjected to in-depth critical study. In these fictions, we find cities that are utopian or dystopian, and the city is habitually a liminal psycho-geographical topos; a metaphysical space which converges with the consciousness of the urbanite. Often, it is presented as a hyper-realistic metropolitan locale, forming an important site for specific social commentary on contemporary Irish culture and society.

Chris Jenks has commented that the city is 'a magical place', 'but the magic is not evenly distributed. So uneven is the experience of city life that it would not be vexatious to describe the idea of an urban culture as oxymoronic'.[7] This view of the city as an impossible space is problematic, and we counter this approach to the intense flux that the city presents by also considering the material qualities of the urban experience. In light of increased globalisation and connective technology, we cannot expect a homogeneous experience of the city; in fact Mark Gottdiener and Leslie Budd suggest the global city has become de-centred as a result of globalisation: '[t]he new information economy, with its accelerating use of all types of electronic telecommunications, possesses counter tendencies of de-centralisation as well as supporting the growth of new centres, including multi-cantered regional growth'.[8] Regional growth here in Ireland has encouraged the development of smaller cities and larger towns outside of the capital, each with their own unique, and culturally, globally, and economically responsive centres. These cities present the potential for new understandings of not only Irish urban experience but also global urban experience.

In the fictions of Kate O'Brien and Kevin Barry, we move beyond Dublin to encounter representations of cities on the western sea-board that respond to both global and local actualities. Huge cultural, political, economic, and colonial changes over the past 100 years have impacted how the city is experienced and as a result how it develops and evolves. Belfast, for example, is what Caroline Magennis calls a city in 'transition', 'haunted by violence'.[9] The economic boom of the Celtic Tiger also had a major impact on the topography of Ireland, urban and rural. The visual reminders of the subsequent crash remain as tens of thousands of houses still sit abandoned in ghost estates across the country. As Smyth notes, the house building boom during the first decade of the 21st Century 'led to an extreme distortion of established life and work practices'.[10] People could no longer afford to live in Dublin and commuted from towns and cities across the country, creating new urban centres, commuter towns, and a new ethno-geography. The Celtic Tiger and its aftermath had a major impact on traditional notions of 'Irish national identity that obtained during the modern era suffered an extreme assault during the closing decades of the twentieth century and continuing on down to the present day'.[11]

While the changing nature of the Irish city has drawn strong creative impulse and literary response, critical attention has been drawn, in the main, to the connotations that link notions of Irish fiction to the regional and the rural. Arguably, the Irish cultural revival at the beginning of the twentieth century and nationalist legacies leading into the 1980s reinforced traditional values relating to Irish identity as cognate with rural experience. These values have resonated intensely through the range of fictions that have come to represent Irish writing since its modernist heyday. Literary greats such as John McGahern, John B. Keane, and Edna O'Brien produced some of the most unforgettable renderings of rural Catholic Ireland and these have echoed through depictions of Irish literature in English by Irish, British, and American authors alike. Oona Frawley discusses this point in her book *Irish Pastoral*, noting that Yeats and Synge and others involved in the revival often relied on 'sophisticated urban concepts of nature […] that allowed them to engage in idealisations that led to the construction of the idea of the Irish nation as rural, traditional'.[12] Significantly, these constructions were tied to issues over land tenure, and the political agendas of nationalism thus became bound to the city as a site at odds with romantic notions of Irishness. While Synge and other revivalists were aware of this growing dichotomy and worked to disable 'the urban/rural binary current at the time', in particular pointing to the commonality of suffering in rural and urban Ireland and the importance of the shift of nationalist agitation from rural to urban settings at the turn of the twentieth century,[13] Dublin continued to be rendered a zone of moral degeneracy in the broader cultural imagination and was long associated with Imperialist capitalism. Catholic anxieties about the threat posed by the capital to national purity and homogeneity are explored by experimental Irish city writers such as Flann O'Brien, who as Laura Lovejoy demonstrates in Chap. 7, begins to dismantle the contrived relationship between Irish identity, Catholicism, and the rural, expanding on the complexities of Dublin life which opened up more fluid concepts of urban life and Irishness.

'Multipli-cities'

Our purpose in this book is to address the city in Irish fiction as a fluid and multiple space that expands traditional notions of Irish identity as rural and challenges the view that cities are homogenous or singular. All of the chapters in this collection are guided by the common objective of re-evaluating the significance of the urban to Irish experience and exploring how the city speaks to the rapidly changing nature of Irish identity which, in the twentieth and twenty-first centuries, consistently responds to large-scale social, cultural, and economic shifts. The chapters form a cohesive perspective that aligns with postmodern critical approaches to cultural identity, agreeing with Jean-François Lyotard that 'eclecticism is the degree zero of contemporary general culture',[14] and thereby we address the diverse range of representations of the urban that are evident in Irish fiction. Our method allows for the deconstruction of the multiple layers of the ever-changing city and for cultural critique of subjective and plural experiences in urban time-space which results in what Michael Peter Smith calls 'a better purchase on the politics of personal and ethnic identity under our present globalized conditions of existence'.[15] As such, we view the cities of Irish fiction as *multipli-cities*, evolved through the swathes of change that have affected Irish identity since the turn of the twentieth century, impacting how we see our 'selves' as a nation, north and south of the border. From this point, we consider how Dublin and other Irish cities exist for the many who call them home through the kaleidoscopic literary lenses offered by literary fiction.

As such, our perspective on the Irish literary city tallies with the Benjaminian idea of urban phantasmagoria. While acknowledging the peculiarities of the Irish city architecturally—being less a city of glass than a palimpsest of history and economy—we align ourselves with Benjamin's view of the city as a spectacle of shifting images to provide an account of representation of the urban in Irish fiction. This view is particularly fitting in contemporary texts which examine the post-Celtic Tiger economy and its impact on the Irish urban experience more broadly. In the work of Kevin Barry, for example, we encounter a city that is fluid and shape-shifting, both grotesque and fascinating—a dystopian response to societal collapse and crime and pollution resulting from excessive urbanisation.

Beyond Dublin and Limerick, we also offer examinations of 'post'-conflict Belfast and its frequent re-imaginings in relation to globalised and 'new Irish' identities. Our strategy is to analyse a spectrum of Irish urban contexts as depicted in Irish fiction. The range of diverse analyses of city fictions is to demonstrate the multifaceted nature of the city in Irish fiction, and Irish fiction in the city. Through theoretically and contextually informed close readings of selected Irish urban fictions, we examine how the Irish city is constructed as a plural space to mirror the plurality of contemporary Irish identities north and south under three distinct, but complementary, strands.

This collection is structured, not chronologically or by region. Instead, it is shaped around interrelated and complementary thematic principles which encourage lateral reading across the collection as a whole, inviting those who engage with the research to make connections across the range of studies, and to participate in a new discourse and understanding of Irish urban fictions. The three sections of the book each follow a distinct but related thematic approach; the sections speak to each other, but also to a particular feature of the concept of the 'living city'. The first, which explores *the city as experience*, considers the interiority of the city and the relationship between city and subject, to discuss 'belonging' in the city and the intersections of identity construction for the Irish urbanite. The second approach examines *the imagined city* and the frequent queer and uncanny depictions of the city that can be found in dystopian, fantastic, and postmodern urban fictions. The final approach is directed toward the notion of *the city in flux* and plural narratives of the city. Exploring how the city is written, not only in literature but from the perspective of each individual city dweller, the chapters which take these approaches to the city collectively generate a discourse of the Irish city in fiction as 'multipli-city'.

Joyce's Dublin and New Imaginings

Frawley suggests that Joyce's Dublin 'represents an obvious point of departure within the Irish literary tradition that wrote so overwhelmingly of the rural'.[16] While this is certainly true, over-emphasising the point could lead one to overlook key urban writers such as Bram Stoker and Sheridan Le Fanu before him, who chronicled 'old Dublin city', its castle, its landmarks, and its shady winding streets, in works like *The Cock and Anchor* (1845). In many ways Le Fanu established Dublin as a literary Gothic city, and this character has remained important to the capital's urban reputation today, drawing tourists to take part in ghost tours and literary walks of so-called haunted locations. But late Victorian Dublin in fiction was more than its uncanny hauntings, as demonstrated in Somerville and Ross's 'startling socio-political reconstruction of late-Victorian Dublin', which, in *The Real Charlotte*, formulates the city as a romantic city of loss for the young protagonist Francie Fitzpatrick.[17]

However, Joyce's rendering of the Irish capital marks an important juncture in the literary envisioning of the urban. As an intensification of the theme of the city of loss, it becomes a city of paralysis, as Liam Lanigan puts it. Not only this, it is contextualised by Joyce in relation to the wider European, and indeed global, contexts to which it related, culturally and economically at that time. According to Lanigan, 'it is assumed that Paris is viewed as a glamorous European capital and a centre of global culture, while Dublin is regarded as a backwater in which the force of Joyce's intellect could not be contained, and could never be fully developed[18]'. This may be the case and it has certainly impacted our view of the exiled artist. However, it also had further interesting repercussions as Desmond Harding points out: 'Joyce's internationalist vision of Dublin generates powerful epistemic and cultural tropes that reconceive the idea of the modern city as a moral phenomenon in transcultural and trans-historical terms.[19]

Joyce's globalisation of Dublin, combined with his simultaneous shift toward the interiority of the Dubliner, radically altered the way that Ireland and the Irish city could be imagined in literature. This cannot be underestimated, and the significance of this innovation for twentieth- and twenty-first-century Irish fictions has informed not only the motivation behind this book but its structure and design. As such, new analyses of Joyce's Dublin have been the ideal place to begin each of the three sections in this book. Three essays which revise traditional readings of Joyce's urban fiction introduce and situate the thematic direction of the critical discourse in each section and the overall commentary of the volume as it moves from discussing the city as experience, to the city as imaginary, and finally to city as amorphous and plural. As part of our methodology, we deliberately look back at that most iconic version of Irish urban identity, Joyce's Dublin, to explore how this literary city transcends its original modernist contexts and speaks to the construction of real and fictional representations of the Irish city in literary and cultural terms. Exploring the legacies of Joyce's urban imaginary, and equally, literary departures from his textualised city of words, allows us to contribute to a more complete picture of modern Irish literature and Irish urban cultural experience.

Dublin, fictionally and physically mapped in *Ulysses* and *Dubliners*, is an important counterpoint to literary imaginings of Ireland as a rural nation. Joyce, referred to by Seamus Deane as 'the first and greatest of Irish urban writers',[20] created a vision of the Irish city which cast a long literary shadow over twentieth-century urban fiction. Generating a vast amount of literary criticism as well as theoretical re-readings and creative re-writings, Joyce's Dublin has become a city much cherished in the Irish imagination. Alternate Irish urban narratives have somehow not warranted the same depth of scholarly interest, and while a prolific body of criticism exists on Dublin in Joyce's work, we are still left with a lack of understanding of what the city really means to Irish literature as a modern cultural phenomenon.

New Directions in Irish Urban Literary Studies

While there has been some important research undertaken on the city in relation to specific Irish authors such as Joyce and Beckett, relatively little work has been done on the city as an Irish literary trope, or to examine its significance to Irish writing more broadly. Choosing to focus on the genre of fiction, this is the first book to address both the city in Irish literature and the Irish city in literature. Our work opens new avenues for research in Irish literary studies and builds on existing research that includes broad-spectrum studies of urban writing. In such studies, including Tom Herron's *Irish Writing London*, although the focus is elsewhere, important references are made to the Irish city, most prominently, Dublin, and to the Irish diaspora experience of cities such as London, Paris, and New York. Deaglán Ó Donghaile's essay, 'Oscar Wilde's Other London',[21] is just one such example from Herron's book, examining the impact of urban experience on Wilde's writing.

Critical writing on late nineteenth- and early twentieth-century Irish fiction has also drawn upon the importance of the urban/rural urban divide as Michael Begnal demonstrates in his fascinating essay on the city of Galway in Finnegan's Wake[22] and as discussed in the 2012 'Reimagining Ireland Series' by Irene Gilsenan Nordin and Carmen Zamorano Llena. This edited collection is notable for examining how the urban/rural divide has been represented in literature since the end of the nineteenth century with focus on the Celtic Tiger as a transformative event on the Irish cultural landscape.

Our work in this volume is also informed by the ongoing research of Irish studies scholars who have been exploring issues relevant to Irish urban fiction. Echoing Seamus Deane's ever-popular *A Short History of Irish Literature*, which retains a strong focus on Ireland's Catholic and colonial/neo-colonial cultural contexts, Aaron Kelly's *Twentieth-Century Irish Literature*, published in 2008, surveys the literary landscape of the late nineteenth and early twentieth centuries in Ireland, dealing with some of the authors that are a focus in this collection. As a seminal text on Irish literature, it is worth noting that Kelly's study also gives little attention to the city beyond the scope of Joyce and his iconic fictionalisation of Dublin. Kelly, like many other scholars of modern Irish literature, addresses the dominant issues of nationhood, Catholicism, gender, sexuality, and the impact of postmodernism on Irish writing. In his study, however, the urban setting of many of the works studied remains a subtext to the main subject of discussion. With a strong focus on the legacy of the Irish revival on literatures of the north and south, Kelly does open an important avenue for the revision of conventional notions of Irish literature and Irish studies as a research discipline. Following this lead, this volume explores Irish literature and its tacit thematics, but does so through a lens which considers the significance of the Irish urban locale. Consequently, it establishes the city within the main frame of discussion to complement explorations of the many social and cultural issues which are the subject of Irish fiction since the late nineteenth century.

Outside of the field of Irish studies and in the broader context of urban literary studies, Irish fiction is becoming an ever more popular subject for discussion and criticism. Until the publication in 2014 of Kevin McNamara's *Cambridge Companion to the City in Literature*, the main resource on urban literature was Richard Lehan's *The City in Literature*, which provides an almost encyclopaedic account of the emergence of the Western idea of the city in literature, spanning genres such as the Gothic and fantasy writing, disciplines such as architecture, urban studies and literary criticism, and literary periods from Romanticism to postmodernism. Coinciding with McNamara's book was a revived interest in the urban in literature as a site of simultaneous order and chaos which reflects social and cultural change. Academic conferences such as 'Cityscapes: Media Textualities and Urban Visions', which took place in 2015 at York St John University, and the 2016 Summer School 'The City in Literature' hosted by The International Association for the Study of Irish Literature testify to the current growing interest in urban literature and urban Irish literature in particular. Similarly, academic networks including the Association for Literary Urban Studies have explored city literatures in a variety of contexts showcasing work on Irish writers such as Dermot Bolger, Tana French, and Roddy Doyle. This demonstrates that much important groundwork has been done on Irish city writers and interest in the urban Irish literature is steadily increasing.

According to recent research published in *The Journal of Urban Irish Studies*, the urbanisation of the Irish population is again steadily on the rise. The Central Statistics Office, Ireland, estimates that the Greater Dublin Area will see its population increase by just over 400,000 by 2031, anticipating a significant return of internal migration patterns. These increases are paralleled by a growing recognition of the importance of urban-related issues in Irish society. Not only are these issues relevant to the fora of politics and urban planning, culturally, there has been a proliferation of literary and artistic responses to changing urban identity. Folkloristics, anthropology, psychogeography, and mytho-geography have informed art, curatorial, and design in terms of their practical and theoretical engagements with the Irish city, and with the rise of literary tourism, there are many new and interesting avenues of research which can investigate the impact of the urban on contemporary Irish literature.

Through the lens of literary criticism, this volume connects the tangible socio-political issues of Irish urban life to the abstract, metaphorical, and symbolic significance of the Irish city. Observing utopian and dystopian representations of the Irish city in fiction of the early twentieth century and a return to Romantic notions of city life in more recent writing, it presents the range of city experiences which have remained important to the development of Irish fiction in the last century. Rather than examining literature set in Irish cities on a politico-geographical basis, the collection is focused not on location but on the concept of the urban and what it means to a variety of Irish authors and contexts. Not wishing to subscribe to chronological or literary-historical orderings, we instead opt to draw attention to the city as site, space, and place in relation to narratives of identity and culture and the literary responses and styles that emerge from investigations into these.

Cities of Change

In this regard, our book takes its cue from Jeremy Tambling's recent collection, *The Palgrave Handbook of Literature and the City*. Viewing the city as something 'amorphous' which resists representation and its possibility,[23] we approach the city as a space at odds with nationalist visions of unity and homogeneity. As Tambling states, the city is dysfunctional in its relationship with the imagined community because it provides a sustained challenge to national consensus and participates in global economics and culture.[24] As such, there is no single history of the city to be told. We investigate Irish urban fictions for their endorsement of this idea. The city in Irish fiction, as we see it, resists mapping, resists singular narrative, and is always in a process of change. In Tambling's book, Jason Finch provides some interesting insights into literary writing as it relates to modern urban theory. Finch suggests that:

> literary writing by Joyce, Woolf and Kafka treats city life in a different way from previously; not in terms of the narrative arc of a life in which one can succeed or fail on moving to the city in a time of uncertainty, as so often in

> nineteenth century urban fiction ... but in terms of moments and chance happenings, inconsequential in themselves but, in their immensely multiple totality, composing city life as experience rather than biography.[25]

The manner in which city life is treated in Irish fiction is also not governed by plot or a sense of narrative ordered by fixed notions of time and space. As a fleeting locale that is simultaneously interior and exterior, it is frequently presented as montage, phantasmagoria, and dreamscape. Notions of Self in relation to the city are radically compromised in this regard. For Kate O'Brien, as much as for Ciaran Carson, the city exceeds the grasp of realistic description and exists in the mind and in memory as intimately bound to phantasies of personal identity.

The de-centred urban space, which revolves around the city dweller and the city visitor, encompassing notions of home as well as the unhomely, and filtering through constructions of self, other, and the crowd, is frequently imagined in Irish fiction as a revenant space. Many of the chapters here highlight how this revenant urban spatiality in Irish fictions is intimately connected to site-specific narrative. And it is in these narratives that the physical/material and the folkloric collide in the recurrence of the past in the present. Ivan Chtcheglov claims that '[a]ll cities are geological; you cannot take three steps without encountering ghosts bearing all the prestige of their legends. We move within a closed landscape whose landmarks constantly draw us towards the past'.[26] This is true of the Irish city and is revealed in urban Irish fiction. The impact of the city on Irish literature is often connected to its haunted and doubled spaces. Arnold Hauser argued that this aspect of the urban was the foundation for the emergence of both the modernist self and impressionism in art and literature.[27] In an Irish context, impressionist and modernist representations were indeed urban phenomena, and the landscape quality of the city came to be depicted through the subjective perspective and stream of consciousness of the city walker. As literary styles changed over time with returns to popular genres along the way, the impressionist vision of the city as an ephemeral space lingered and the city itself became a metaphor through which to explore not only urban themes such as alienation, isolation, and identity-mapping but also those themes important to the Irish setting, including the colonial experience and Catholic identities. In this way, the Irish city in fiction is a unique space. Its everyday contexts stand apart from the cities in British, European, and American contexts, as part of the legacy brought to bear by its colonial heritage, its neo-colonial institutions, and its contemporary post-conflict status.

This volume proposes to establish an up and coming field of research which specifically traverses the disciplines of urban literary studies and Irish studies bringing the study of Irish fiction into discussions of urban literature in a global context.[28] Irish urban literary studies, in our case examined through a focus on fiction, brings with it research interventions relevant to postcolonialism, gender studies and theory, modernism and postmodernism, the Gothic and genre fictions, minority experience in literature, psychogeographies, spatial theory, and social studies. In this context and drawing from multiple perspectives and approaches, we explore the city as a cultural marker, examining what it means to be an Irish urbanite and how this finds cultural representation within the broader frameworks of national and cultural identities. From this range of perspectives, this collection looks at representations of the Irish city in fiction as well as the city in Irish fiction. Writers and characters in these fictions engage with Irish places, fictional and real, and the cultural implications of the specifics of representation are important. The chapters in this volume examine how Irish urbanites see themselves within their own urban spaces and how narratives of identity are conjoined with narratives of place. In doing so they ask how does the rural Irish literary tradition impact on their experience of the city? How do urban Irish locations differ from recurring images of the city in literature? And does the unique cultural and historical past and present of the island of Ireland create contrasting urban experiences north and south of the border? These are questions specific to Irish urban fictions. They challenge us to think about important issues relevant to the literary urban landscape. McNamara suggests that literary forms have been the 'building blocks of [...] collective identity for millennia'.[29] If so, then what does the collective cultural identity of the urban suggest about Ireland? And how has urban literature contributed to the development of this collective identity?

A Multilateral Perspective

The chapters in this book approach the city and the urban spaces that unfold within and around it, by drawing the concept of 'Irish fiction' away from notions of the rural and the regional. The collection's vision expands the idea of the city in Irish writing to reveal a host of utopian, dystopian, and heterotopic textual spaces that frequently converge with the consciousness of the city dweller. In doing so, it forms a perspective which opens up the urban in Irish fiction as an important site for social commentary on contemporary Irish culture and society. Each of the three sections builds on these notions, and the chapters therein combine to form a critical introduction to Irish urban fictions. They engage with current discussions in Irish Studies and Urban Literary Studies to develop a critical commentary driven by individual literary analyses. Examining the crucial relationship between the city and literature, they progress our understanding of modern Irish fiction and Irish literature more generally, and of the Irish city itself. Along with analysis from the broader field of urban studies, this collection adds to the conversation of the city experience in a globalised world.

The collection begins discussion and prompts enquiry around three tacit themes relevant to Irish urban fictions. It explores a range of experiential urban contexts which determine the relationship between the city and the subject. It examines and discusses a wide and diverse range of authors, including James Joyce, Roddy Doyle, Rosemary Jenkinson, Ciaran Carson, Glenn Patterson, John Banville, Flann O'Brien, Kate O'Brien, Hugo Hamilton, and Kevin Barry. Brought together with the aim of generating a cohesive perspective on the significance of Irish urban fictions, they stimulate new research in the area and combine to offer a multilateral perspective on the Irish city in literature and the versions of Irish subjectivity and spatiality unique to these literary spaces. Representations of the Irish city vary from converging dystopian and utopian modernist cityscapes to the more ephemeral and fragmented spaces imagined in late modernist and postmodernist fictions. Across these fictional cities, we find hints of nostalgia and romanticism and narrated urban experiences of de-centred belonging. As a complex of multiple intertwining and clashing spaces and stories, the texture of the city in Irish fiction is revealed as shifting and fluid, and it opens a multitude of possibilities and perspectives for the Irish city dweller, as it does for the Irish city writer.

Part I, 'Whose City Is It Anyway? *The City as Experience*', offers four chapters that deal with the concepts of belonging and home from a perspective that views the city as a subjective and interior literary landscape. Eva Roa White's Chap. 2 considers this question in relation to Dublin as it has been experienced subjectively by the characters of Joyce's *Dubliners* and Doyle's *The Deportees* and *Bullfighting*, which, she argues, stand in their own time capsules of life in Dublin in the twentieth and twenty-first centuries. White contends that Joyce's 'ownership' of Dublin must be revisited through the literary lenses of authors such as Doyle who reflect contemporary and cosmopolitan urban sensibilities. Interested in the multicultural dimensions of contemporary Dublin and how this is represented in literature, this chapter offers a discussion that is also keenly aware of the haunting nature of Joyce's city which is inseparable from 'new Dublin'.

Chapter 3 explores the relationship between another iconic early twentieth-century writer, Kate O'Brien, and her city of Limerick. Margaret O'Neill contends that the real city of Limerick influenced O'Brien's perspectives on identity in flux during this crucial period of literary modernism in Ireland. Focused on the prevalence of the city in O'Brien's work, and its function as a *locus standi* for the writer, O'Neill points to how the tangible nature of the city with its spires and market squares, slips through the author's grasp, existing as an abstract and conceptual space through which her literary investigations can be explored. She does this, with a historian's sensibility, exploring the social changes and tensions of modernity, the creation of the Irish Catholic middle class, and the position of the woman and artist in modern Irish society.

While O'Neill examines the concept of the native and her city, Molly Ferguson, in Chap. 4, discusses the otherness of the city from the perspective of the 'new Irish'. In this chapter, Ferguson explores the city as 'a constellation of affective public and private spaces—the home, the pub, the court, the pier, the street, and the church' and follows the meanderings of Hugo Hamilton's migrant character Vid as he encounters a city of national monuments which are alienating to him as a stranger to the city. Using Jacques Derrida's concept of hospitality, Ferguson explores the novel's representation of the challenge to Ireland's reputation for hospitality by revealing an underside to the city of Dublin, run by the paid labour of an invisible underclass.

This view of the city in relation to narratives of inclusion and exclusion is echoed in Tim Keane's chapter which follows. Through an analysis of Ciaran Carson's *The Star Factory*, Keane discusses how Carson presents Belfast as a heterogeneous and apparently unreal city. Taking his cue from Carson's 'ongoing, fractious epic that is Belfast', Keane considers the city as a disorienting locus of deceptive images, commodity fetishes, multifaceted icons and symbols, and hybrid interior-exterior spaces. Keane suggests, with reference to Benjamin's concept of the aura, that the dreamlike components which define the urban experience of Belfast confirm Calvino's claim that cities emerge from the interplay of our deepest fears and desires. Belfast, like many literary cities, is thus a sort of unconscious realm accessible through its own unique language; constructed from signs, symbols, and personal mappings.

The central part of the volume, entitled 'Disturbing Phantasies and the Uncanny City', follows the surreal and disjointed image of the city—identified earlier in Carson's Belfast—through an exploration of manifestations of the city in Irish fiction as a fantastic and frequently uncanny environment. Considering the delicate relationship between order and disorder in the city, these chapters engage with a fear identified by Lieven Ameel, who regards the city as the opposite of utopia: that 'an imminent end-time could upend the social and cultural fabric of humankind'.[30]

Chapter 6 explores this unsettling and often anxious milieu of the Irish city in fiction in relation to the notion of the city as a strange, almost dystopian imaginary. Its author, Quyen Nguyen reminds us that while Joyce's fictionalised urban space has been a preoccupation of critics of all persuasions: cartographical, psychoanalytic, sociological, and postcolonial, it should also be understood as a discourse between the city and its 'user', Leopold Bloom. Because the city is 'the writing' and the citizen 'its reader', Joyce's Dublin opens a strange in-between space with possibilities for unusual subjective urban literary experiences.

Laura Lovejoy, in Chap. 7: 'Urban Degeneracy and the Free State in Flann O'Brien's *At Swim-Two-Birds*', then goes on to consider themes of morality and degeneracy in relation to the fluidity of O'Brien's fictional Dublin. As a challenge to the 'Catholicised vision of Ireland's post-independence purity', Lovejoy argues that O'Brien's Dublin is presented through a moral topography which is anti-realist and formed through a series of 'intra-narratives'. The degeneracy of the city, she claims, stands as a 'spatial focal point of Free State anxieties surrounding urban-centred immorality'.

Moving from moral degeneracy to urban crime, Martyn Colebrook discusses the 'urban' of 'disturbance' in Chap. 8 through a close reading of Kevin Barry's *City of Bohane*. Focusing on the novel's creation of a decaying post-industrial Gothic cityscape, Colebrook investigates how the representation of the city of Bohane, links Barry's fiction to the genre of Irish Gothic. As he reminds us, it is '[a] haunted Bohane' which 'reveals the manifest layers of voices, the polyphonic and echolalic city that will be forever in debt to its own ghosts'.

Chapter 9, by Neil Murphy, turns toward the urban fictions of another writer closely associated with Irish Gothic, John Banville. Murphy reminds us that the urban settings of many of Banville's novels, including Dublin and Rosslare, function as a mirror to the language-obsessed subjective consciousnesses that dominate his fictions. Murphy compares Banville's cities to those of Calvino and explores how the texture of his aestheticised urban realities reveals the urban as a site which insists upon its own otherness as a ghost of itself—a strange mirror imaginary. Together, the chapters in this central part of the book provide an essential discussion of the fractured and refracted versions of the city that resonate in fiction, but also beyond it in the imaginary that makes up the city in and of itself. They point to how literature reveals the true nature of the urban as an uncanny and multiple environment that exists only in its interactions with its inhabitants and visitors.

With a focus on multi-layered and multi-dimensional nature of the city, often the locus of postmodern urban fictions, the final section, 'Cities of Change: Re-writing the City', explores how plural and shifting narratives can create the city. Signalled by Murphy's discussion of Banville in the previous section, the three chapters in this part explore textuality of the city and the relationship between the city, the imagination, and language. Nick Bentley has argued that 'the complexity of the contemporary urban space is rendered in the postmodern novel through a pluralisation of space, time, and social discourse'.[31] In this section we see the move from the spectrality of the modernist city to the postmodern city of signifiers and fluid identity in the layered realities of the contemporary living city.

This is firstly explored by Nikhil Gupta in Chap. 10, entitled 'The Haunted Dublin of *Ulysses*: Two Modes of Time in the Second City of the Empire'. Gupta is particularly interested in the presence of ghosts in Joyce's Dublin and in the 'Wandering Rocks' episode of *Ulysses*. He discusses how these spectres function to underscore the dual nature of life and time in the city. Examining Joyce's Dublin in relation to the idea of a national community, this chapter discusses the temporal dissonance of the national capital as a city looking to the future, but haunted by the past.

Discussing a city also frequently seen as a contested site trapped between narratives of past and future, our penultimate chapter, 'It's only history': Belfast in Rosemary Jenkinson's Short Fiction investigates Belfast as 'a place apart'. In this chapter, Dawn Miranda Sherratt-Bado reads Jenkinson's short stories considering Colin Graham's point that in Northern Irish cities 'peace was bought at the price of dissociation rather than consociation'. The theme of alienation runs through the urban fictions discussed in this chapter, and the author significantly explores Jenkinson's visualisation of the contemporary city as a commentary on the commercialisation of Belfast and its history.

The final chapter (Chap. 12) in the collection is also focused on Belfast, but this time in relation to the novels of Glenn Patterson. Terry Phillips repeats an important commentary in this chapter on how contemporary studies of the urban highlight the very contradictory nature of the city as sites of simultaneous order and freedom. Reminding us that the city is a shared space, impermanent, and fluid, Phillips reads Patterson's urban novels (*Burning Your Own*, *The International*, and *Number 5*) for the multiple shared identities that they construct in relation to gender, nationality, ethnic affiliation, and class. The power relations of cities are brought into focus as Phillips discusses Patterson's broader literary interest in the city's history and its landscapes.

As you read through the chapters in the collection, you will move through many cities and their equally numerous representations. These chapters, like the cities they discuss, speak to Roland Barthes' conception of the city as a discourse to be engaged by the city dweller or city walker. The city in literature is much like the architectural reality of the city. Our understanding of it must shift in relation to each subject, or character, that perceives it.[32] The city is subjective. It is fragmented and impressionistic: to walk the city is to read it. Barthes noted that there exists 'a conflict between signification and reason, or at least between signification and that calculating reason which wants all the elements of the city to be uniformly recuperated by planning'.[33] Irish urban fiction navigates this gap when it writes of cities of the imagination, and this collection begins a conversation that includes but moves beyond Dublin, and the physical border on the island to approach the discourse of the city in Irish fiction. As such, the analysis herein illustrates how the Irish city and the writers of these cities engage with notions of nationality, identity, history, memory, and our globalised future.

In terms of literary and cultural criticism, it is not possible to offer a complete picture of the urban in Irish literature. The city is never a static entity. Instead, we expect that the range of authors and cities discussed in this book begin to enliven the literary image of the Irish city as a place that has evolved through Romantic, modernist, and postmodern ideas while retaining its own distinct characteristics. As both a physical and a metaphysical space, the Irish city in fiction reflects the sublime and the seedy; the uncanny and the domestic; the past and the future. The body of work presented in this volume reveals an important new critical cognisance of the significance of the city to Irish literature. As much as literature has been important to the cultural identity of Irish cities, north and south, the city too has remained central to the Irish literary imagination. We envisage that the collection as a discursive tool can broaden our understanding of Irish fiction in the late nineteenth and twentieth centuries and aid new research into the importance of the Irish city to our contemporary literary cultures.

Notes

1. www.dublincityofliterature.ie.
2. As Joseph Valente argues in his introduction to Eire/Ireland 45, 1&2: 'The urban experience has indeed been comparatively underappreciated' in studies of Irish literature'. Frawley 2010, p. 10.
3. For more on the global city, see Harvey, D. 2000. Spaces of Hope. Edinburgh: Edinburgh University Press.
4. It should be noted that 'city' is used here as a term to denote an urban location with a measured population, functioning as a municipal centre. 'Urban' references the geographical location of dense and organised population settlement and can be used to designate a large town or city. And finally, 'metropolitan' refers to larger cities and their less densely populated suburban regions which together function as a socio-cultural, political, and economic centre.
5. Smyth, Gerry. 'The Right to the City: Re-presentations of Dublin in Contemporary Irish Fiction'. In Contemporary Irish Fiction: Themes, Tropes, Theories, edited by Liam Harte and Michael Parker. 13–35. Basingstoke: Palgrave Macmillan, 2000, p. 14.
6. Ibid.
7. Jenks 2004, p. 1.
8. Gottdiener & Budd 2005, p. 44.
9. Magennis 2016, p. 219.
10. Smyth, Gerry. 'Irish National Identity After the Celtic Tiger'. Estudios Irlandeses 7, no. 7 (2012): p. 133.
11. Ibid., pp. 134–135.
12. Frawley 2005, p. 53.
13. For more on this, see Mathews, PJ, *Revival: The Abbey Theatre, Sinn Féin, the Gaelic League, and the Co-operative Movement*. Cork University Press: Cork. 2003, p. 75.
14. Lyotard 1992, p. 8.
15. Smith 1994, p. 494.
16. Frawley, p. 106.
17. Harding, Desmond. Writing the City: Urban Visions and Literary Modernism. London: Routledge, 2003. pp. 49–50.
18. Lanigan, Liam. *James Joyce, Urban Planning and Irish Modernism: Dublins of the Future*. Palgrave. London, 2014, p. 1.

19. Harding 2003, xi.
20. Deane 1994, p. 180.
21. In Herron 2012, Vol. I.
22. See Begnal 2002.
23. Tambling 2016, p. viii.
24. Ibid., p. x.
25. Finch, in Tambling 2016, p. 36.
26. Chtcheglov 1953, p. 58.
27. Hauser 1999, p. 168.
28. While this book is the first of this kind to critically examine the representation of the city in Irish fiction, we must acknowledge Andrew Kincaid's *Postcolonial Dublin* as a forerunner for its investigations into the Irish city in cultural and economic terms. Kincaid brings the city of Dublin back into important political and cultural discussions of modern Irish identity. With a specific focus on architectural and planning history, he reignites debates about the formation and consolidation of the modern Irish State by problematising the prevailing image of Ireland as a rural idyll often constructed through art and literature. Kincaid 2006, p. xvi.
29. McNamara 2014, p. 1.
30. Ameel, in Tambling 2016, p. 785.
31. Bentley, Nick 'Postmodern Cities'. In The Cambridge Companion to The City in Literature, 200–215. Edited by Kevin McNamara. Cambridge: Cambridge University Press, 2014, pp. 175–188.
32. See Barthes, Roland. 'Semiology and Urbanism'. In *The Semiotic Challenge*, translated by Richard Howard, 191–201. Oxford: Basil Blackwell, 1988, pp. 191–201.
33. Ibid., 194.

References

Barthes, Roland. 'Semiology and Urbanism.' In *The Semiotic Challenge*, trans. Richard Howard, 191–201. Oxford: Basil Blackwell, 1988.

Begnal, Michael H., ed. *Joyce and the City: The Significance of Place*. New York: Syracuse University Press, 2002.

Benjamin, Walter. *Selected Writings 1938–40*. Boston, MA: Harvard University Press, 2003.

Bentley, Nick. *The Arcades Project*. Boston: Harvard/Belknap, 1999.

———. 'The Work of Art in the Age of Mechanical Reproduction.' In *Illuminations*. New York: Schocken Books, 2007.

———. 'Postmodern Cities.' In *The Cambridge Companion to the City in Literature*, ed. Kevin McNamara, 175–188. Cambridge: Cambridge University Press, 2014.

———. *One-Way Street*. Boston Harvard Belknap, 2016.

Calvino, Italo. *Invisible Cities*. San Diego, CA: Harcourt Brace Jovanovich, 1978.
Chtcheglov, Ivan. 'Formulary for a New Urbanism.' *The Internationale Situationniste*, No. 1, June, 1958.
CSO Ireland Figures. http://www.cso.ie/en/releasesandpublications/er/rpp/regionalpopulationprojections2016-2031/.
Deane, Seamus. *A Short History of Irish Literature*. Indiana: Notre Dame University Press, 1994.
Foley, R., and J. Sweeney. 'Introduction.' *The Journal of Irish Urban Studies* 7–9 (2008–2010): 1–3.
Frawley, Oona. *Irish Pastoral: Nostalgia and Twentieth-Century Irish Literature*. Dublin: Irish Academic Press, 2005.
Gottdiener, Mark, and Leslie Budd. *Key Concepts in Urban Studies*. London: SAGE Publications Ltd, 2005.
Hauser, Arnold. *The Social History of Art*. London: Routledge, 1999.
Herron, Tom. *Irish Writing London Vols. I & II*. London: A&C Black. 2012.
Jenks, Chris. *Urban Culture: Critical Concepts in Literary and Cultural Studies*. London: Routledge, 2004.
Kincaid, Andrew. *Postcolonial Dublin*. Minneapolis: University of Minnesota Press, 2006.
Lehan, Richard. *Literature and the City: An Intellectual and Cultural History*. Berkeley, CA: University of California Press, 1998.
Lyotard, Jean-Francois. *The Postmodern Condition Explained*. 3rd ed. Minneapolis: University of Minneapolis Press, 1992.
Magennis, Caroline. '"That's Not So Comfortable for You, Is It?": The Spectre of Misogyny in *The Fall*.' In *The Body in Pain in Irish Literature and Culture. New Directions in Irish and Irish American Literature*, ed. F. Dillane, N. McAreavey, and E. Pine, 217–234. Basingstoke: Palgrave Macmillan, 2016.
Mathews, P.J. *Revival: The Abbey Theatre, Sinn Féin, the Gaelic League, and the Co-operative Movement*. Cork: Cork University Press, 2003.
McNamara, Kevin. *The Cambridge Companion to the City in Literature*. Cambridge: Cambridge University Press, 2014.
Smith, Michael Peter. 'Postmodernism, Urban Ethnography, and the New Social Space of Ethnic Identity.' *Theory and Society* 21, no. 4 (1992): 493–531.
Smyth, Gerry. 'The Right to the City: Re-presentations of Dublin in Contemporary Irish Fiction.' In *Contemporary Irish Fiction: Themes, Tropes, Theories*, ed. Liam Harte and Michael Parker, 13–35. Basingstoke: Palgrave Macmillan, 2000.
———. 'Irish National Identity After the Celtic Tiger.' *Estudios Irlandeses* 7, no. 7 (2012): 132–137.
Tambling, Jeremy. *The Palgrave Handbook of Literature and the City*. Basingstoke: Palgrave Macmillan, 2016.
Valente, Joseph. 'Urban Ireland.' *Eire/Ireland* 45, no. 1&2 (2010).

PART I

Whose City Is It Anyway? *The City as Experience*

CHAPTER 2

Whose Dublin Is It Anyway? Joyce, Doyle, and the City

Eva Roa White

The name of James Joyce is entwined with Irish fiction to the point that it is almost impossible to discuss any type of Irish urban literature without a reference to his works featuring Dublin. Andrew Kincaid connects Joyce to contemporary Irish noir by stating that novels in this genre harken back to Joyce's *Dubliners* because they too deal with one of Joyce's central theme: 'the tension between a parochial Irish urbanism and the desire for a metropolitan culture that has its roots elsewhere.'[1] Though this tension is still very much present in Dublin today, it is no longer present in terms of this desire. Rather, the tension comes from the actual mixing of the local (native Irish culture) and the global (multicultural immigrant culture) in the city itself through the inward migration produced by the brief economic boom of the Celtic Tiger. This new 'glocal' reality that adapts the local to these global influences is documented in the works of Roddy Doyle, particularly his contributions to immigrant magazine *Metro Èireann* chronicling glocal encounters and resulting hybridisations, which he also published as the short story collection *The Deportees* in 2007. It is a testament to Joyce's grip on the Irish psyche and literary scene that in spite of the new visions of Dublin offered by Doyle and others, the disconnect between what was 'old' Dublin and what is 'new' Dublin persists.

E. R. White (✉)
English Department, Indiana University Kokomo, Kokomo, IN, USA
e-mail: evawhite@iuk.edu

M. Beville, D. Flynn (eds.), *Irish Urban Fictions*, Literary Urban Studies, https://doi.org/10.1007/978-3-319-98322-6_2

The fact that every year on June 16, Bloomsday commemorates Joyce's novel *Ulysses* by recreating the events of Leopold Bloom's one-day urban odyssey around Dublin is a case in point. This international literary pilgrimage reasserts Joyce's ownership of.the city and Dublin's dependence on cultural tourism. The centenaries of the writing and publication of *Dubliners* celebrated in 2014 further buttress Joyce's standing as symbol of Dublin and Irishness, both at home and abroad. The initial cultural appropriation of Joyce to construct a cosmopolitan image of Ireland can certainly be understood. After all, in *Ulysses*, Joyce gives us Leopold Bloom, a hybrid (Irish, Hungarian, Jew) character who looks to Europe, not Ireland, for cultural identity. Now that Ireland is more cosmopolitan with a large population of inward migrants from different parts of the world, not just Europe, Joyce's ownership of Dublin needs to be revisited in terms of living authors such as Doyle, who truly chronicle the new Dublin in its present multicultural identity and offer an alternative to viewing Ireland solely through the Joycean lens. Though Joyce uses internationalism in his works, unlike Doyle, it is one that presents Dublin as something to escape rather than embrace. I suggest that both authors have created their own psycho-geography of Dublin a hundred years apart, producing an extension of Kincaid's term, 'urban memoir,' to include not just each author's relationship to the city but his very own vision of Dublin that reflects and crystallises the city's identity at that particular time in history. As Kincaid explains: 'What is clear is that an altered and altering landscape necessitates its own literary form: the urban memoir—a genre as individualistic as the personal coming of age story and as rooted in the physical environment as theories of geographical and psychological change.'[2]

The result is a memoir of the city that reflects the authors' respective attitudes towards Dublin. Indeed, as Margaret Hallissy shrewdly points out:

> Literature reflects not so much history itself as the writer's viewpoint on history. While any view-point might be, in James Joyce's description of Irish art in *Ulysses*, a "cracked lookingglass," to the reader of literature, the cracks reflect their own sort of light. To extend Joyce's metaphor, it is now time to examine the structure upon which this mirror depends.[3]

In this chapter, I build on Michael Malouf's psychoanalytical analysis of the Irish ten-pound note, with the help of Derrida's concept of hauntology and his spectre to problematise Joyce's relationship with Dublin and his psychological hold on the city. I propose that Joyce's ownership of the capital is one that needs to be re-evaluated, in light of the city's new multicultural identity, in terms of the new voices, such as Doyle's, who chronicle the new Dublin and free it from the hold of an author who, much as a Derridean spectre, inhabits the city through what Derrida terms a 'spectral moment':

> A spectral moment, a moment that no longer belongs to time, if one understands by this word the linking of modalized presents (past present, actual present: "now", future present). We are questioning in this instant, we are asking ourselves about this instant that is not docile to time, at least to what we call time. Furtive and untimely, the apparition of the specter does not belong to that time, it does not give time, not that one: "Enter the ghost, exit the ghost, re-enter the ghost." (Hamlet)[4]

Joyce as Spectre

Though Joyce uprooted himself, he never truly left Dublin, in that he never was free of the city or family he left behind. By exiling himself, Joyce turned himself into a ghost forever haunting Dublin. Derrida's hauntology theory, which he introduces in *Spectres of Marx* and relates to his concept of *différance*, is helpful in analysing this phenomenon because 'the true origin of a sign is always spectral. More broadly speaking, Derrida's *hauntologie*—a combination of the Heideggerian and Freudian uncanny—concerns the notion that living in the present is always affected by (ghosts of) the past.'[5] As Colin Davis states, 'hauntology supplants its near-homonym ontology, replacing the priority of being and presence with the figure of the ghost as that which is neither present nor absent, neither dead nor alive.'[6] This relates to Derrida's concept of *différance*, in that the absence/presence of the revenant or ghost consists of traces of the old in the new and vice versa. In this way, Joyce inhabits hauntology itself, in that he is the obsessive spectre that keeps appearing in the present, a trace of the old Dublin that cannot be shaken off, as it exists both in the past and present through what Derrida calls 'disarticulated, dislocated' time.[7] Indeed, this dis-location adds to the deconstruction of time as in the spectre appearing out of sequence. In addition, it is not just the author who becomes a spectral presence but also his works. As Derrida states, 'A masterpiece always moves, by definition, in the manner of a ghost.'[8] In Joyce's case, his novels are also spectral, in that they too haunt the new

Dublin, not only through events such as Bloomsday, discussed earlier, but also as traces in the majority of literature about Dublin.

Hauntology works in two ways with Joyce: He is haunted by Dublin and in turn haunts Dublin, transforming himself into a ghost or living dead of sorts. Joyce's coping mechanism is to own the city he left behind by mummifying it in his writing to help his Dublin withstand the onslaught of both time and space. Joyce feels the psychological need of the exile (even if it is self-exile) to stay connected to his motherland and cannot avoid the backward look to the past that typifies this separation. It is a way to belong to a community that is now mediated by time and distance. To the point that Joyce needs help in reconstituting the physical geography, if not the psycho-geography of Dublin. Joyce's is an Ireland of the mind, his Dublin frozen in time and painstakingly reconstructed street by street from memory and details provided by friends and relatives still living there. As Richard Ellmann explains:

> Joyce, although he transformed those places into words, did not invent them. He said, "He is a very bold man who dares to alter in the presentment, still more to deform whatever he has seen and heard" (May 5, 1906). This was in connection with the book *Dubliners*. He was always trying to verify details of the city which lay almost a thousand miles from the table at which he was writing about it … He was at once dependent upon the real and superior to it.[9]

His compulsive gathering of every detail every street and landmark (even down to ticket stubs) is reminiscent of a lover who tries to gather mementos of the loved one left behind. Johnson refers to Joyce's well-known comment to Frank Bogen as to how Dublin could be reconstructed from his novel *Ulysses*. Joyce's meticulous virtual recreation of Dublin succeeded in imposing his vision of Dublin and established his ownership of the city on which he was fixated.

Michael Malouf's analysis of the now-defunct Irish ten-pound note illustrating Joyce's popular image as what Malouf terms an 'official, tourist-board approved, Irish historical figure'[10] provides good evidence of Joyce's dominance of the city. Assuredly, what began as an effort to legitimise Ireland's cosmopolitan and European identity has resulted in an industry based on cultural tourism that fetishises Joyce and transforms him into a spectral author who haunts the city. As Luke Gibbons notes, 'specters emanate from the market and technology as well as spiritual revivals, and capitalist modernity is no less subject to illusory manifestations than its superstitious counterparts.'[11] Joyce's commodification only reasserts his spectral occupation of the city. The note, in use from 1993 to 1999, features, on one side, a domesticated, if not sanitised portrait of Joyce that most of his contemporaries, familiar with his contentious relationship with the city and its publishers, would not recognise. Smiling benevolently down on Dublin and environs, Joyce floats over Dublin Bay, with Howth Hill behind him, much as a ghost would. The other side of the note displays a statue of Anna Livia's head, the anthropomorphic representation of the river Liffey in *Finnegans Wake*, side by side with the first sentence of this novel, ratified by James Joyce's signature, all floating over a map of Dublin, thus confirming what had already made very clear by Joyce's portrait on the first side of the note: that Dublin belongs to Joyce.

Another example of Joyce's works attaining the status of national and cultural symbols is the placement of the statue of Anna Livia on O'Connell Street, as part of the 1988 Dublin Millennium celebrations. This was not a permanent location however. 'The Floozie in the Jacuzzi,' as it was affectionately nicknamed, was relocated in 2001 to make room for a redevelopment of the area as part of the building boom of the Celtic Tiger. After ten years in storage, the piece was finally placed in the quieter Croppies Memorial Park opposite the Guinness brewery.[12] It is telling that the fountain was replaced in 2002 by a more fitting symbol for the era of the Celtic Tiger: the Spire of Dublin. Also known as 'The Monument of Light' (*An Túr Solais*) and by nicknames such as 'The Spike,' it was chosen by an international committee to symbolise the new relationship between art and technology as well as that of Ireland with its new global family. Rising above the rooftops of Dublin, the Spire reaches well beyond the city pointing to the future and to globalisation. As a result, this symbol is more representative of Doyle's Dublin than of Joyce's. However, despite these unequivocal signs of changing times, Joyce's hold on the city has not diminished. His ghost, or Derridean spectre, still floats over Dublin, 'neither dead nor alive,' but 'hovering between life and death, presence and absence, making established certainties vacillate'[13] and keeping secrets. These secrets are both of a political and personal nature. As Malouf notes, 'It is ironic that Joyce, who created in the Circe chapter of *Ulysses* a critical demystification of commodity fetishism, has become such a magical object himself,'[14] enough to cover the divided nature of the country and associate Ireland with Europe. Malouf continues: 'the figure of Joyce mediates for a nationalist identity refashioning itself for Europe.'[15] The personal aspect deals with how his representation as national symbol hides how much Joyce hated Dublin, his absence during the defining moments of Irish history, and how *Ulysses* was banned in Ireland.

Though Derrida encourages us to speak with the spectre, it is without expectation of any revelation.[16] But the unveiling of secrets is exactly what psychoanalysts Abraham and Torok, whose work preceded and inspired Derrida's hauntology, demand from their own phantom, who prevents shameful secrets from coming to light. For Abraham and Torok, 'the phantom is a liar; its effects are designed to mislead the haunted subject and to ensure that its secrets remain shrouded in mystery (*L' Écorce et le noyau*, p. 427).'[17] The figure of Joyce conjured on the Irish ten-pound note is just such a phantom liar preventing his shameful secrets from coming to light, not of this world, but still hovering over it, with his eye on the city that has haunted him during his lifetime perhaps as much as his legacy haunts Dublin after his death.

Ireland's strategy of choosing Joyce as the living dead signifier for its new cosmopolitan Ireland has paid off. The country has now achieved a multicultural status that was not anticipated even by Joyce when he proposed to put Dublin on the global map. The price, however, has been that of a posthumous commodification of the writer that tries to conceal, if not erase, his contentious relationship with his homeland.

Joyce's Self-Exile

Feeling the need to leave the city in 1904 to succeed as a writer, Joyce left Dublin physically, but not psychologically, in that he never was free of the city he abandoned. Jeri Johnson begins her essay 'Literary Geography: Joyce, Woolf and the City' with Virginia Woolf's words, 'a writer's country is a territory within his own brain.'[18] To be sure, a writer's interior geography or inner-scape does determine his or her national and cultural identity. How this territory is constructed is often informed by physical and psychological shifts, or geographical or internal diasporas. The diaspora of the mind is provoked by a need to '(1) to distance oneself from one's country, race, class, family or other aspect of one's birth identity that does not meet one's physical and/or emotional needs; (2) to join another culture with which one feels a connection'[19] to find a place that fits in the world. The result is identity migration, 'Invisible and interior in scope, it is the inner-scape of what I call identity migrants who construct for themselves a grafted identity. These hybrid identities encompass at least two worlds: the world of origin and the adopted world. The result is interior or cultural hybridisation.'[20] According to this definition, Joyce was an unsuccessful identity migrant, in that all the years he lived abroad, although his aspiration was to give Ireland an international voice, he remained fixated on Dublin. Indeed, his physical migration did not translate into a psychological one. While his body inhabited foreign lands, his soul never left Ireland. As Herbert A. Kenny remarks, 'He [Joyce] fled the Catholic Church and Ireland to remain an exile from both but indelibly tinctured by both in the depths of his soul.'[21] In fact, Joyce did not include his new countries, Italy (Trieste), France (Paris), and Switzerland (Zurich) in his literary psycho-geography, but featured only Dublin as the setting for his major works.

One could say that Joyce's coping mechanism was to 'own' the city he left behind by mummifying it in his writing so that his personal Dublin could withstand the onslaught of both time and space. He felt the psychological need of most exiles to stay linked to his motherland to the point that his connection to Dublin was almost physical. Luke Gibbons points out this somatic relationship to the city to explain Joyce's need: 'Joyce may have created the phantom text of Dublin for the same reason that an amputee imagines a phantom limb: to compensate for the pain of loss'[22] of a city that he also felt was part of himself to the point that Gibbons' statement that 'the experience of a phantom limb is far from being an aberration: it is often the absence rather than the occurrence of the phantom limb that requires explanation.'[23]

Joyce's psychic connection was mediated by time and distance, his absence transforming the Dublin of the past into a city of the mind, as well as a memory. Gibbons points out that 'it is difficult not to think of Joyce's almost somatic retention of the irregularities of his home ground, the city of Dublin' in view of Oliver Sacks' statement that 'a phantom is more like a memory than an invention,'[24] a memory that haunted Joyce into making his obsession with Dublin a literary reality. In a way, Joyce's oeuvre is a sort of exorcism of his demons and his own haunting that would allow him to turn the tables and fashion himself into the Derridean spectre of Dublin, dictating his psycho-geography of the city for generations to come.

Joyce's Secrets

Joyce did succeed in imposing his vision of Dublin, to the point that his name has become almost synonymous with that of the city, although the Joyce of the ten-pound banknote is a phantom keeping secrets. While

Joyce did achieve a cult status in Ireland and abroad after the ban on *Ulysses* was lifted, he also was criticised for his lack of concern for his city. According to Kenny:

> The reasons for Joyce's lack of popularity in Dublin at this time were numerous. No book in history has as much of Dublin in it as *Ulysses*, and Joyce himself said, if the city was destroyed, it could be rebuilt from his works. But many Dubliners viewed Joyce as a man destroying the city as surely as British guns. He was sitting on the sidelines scoffing at his country in its hour of anguish and deliverance.[25]

This Joyce is a far cry from the one promoted by Ireland's department of tourism. As Rubén Jarazo Álvarez explains in his study, 'Managing Culture in Ireland: Literary Tourism and James Joyce,' there has been a progressive push towards cultural tourism in general and featuring Joyce since the 1980s. Not only has the James Joyce Centre in Dublin thrived since its opening in 1996, but it also has achieved an international reputation that feeds not only visits to the centre, but to Bloomsday, the other 'epicenter for the management of James Joyce, whose board includes members of the Department of Tourism.'[26]

Joyce's popularity remains high today in literary and academic circles as well. His prediction that his writing would 'keep the critics busy for three hundred years'[27] has come to pass. The centenaries of the writing and publication of *Dubliners,* marked by *The New Dubliners* edited by Oona Frawley in 2005, and commemorated in *Dubliners 100,* edited by Thomas Martin in 2014, provide us with a clear example of Joyce's ascendancy over the city's literary scene. Although these new collections offer portrayals of Dublin in the twenty-first century, they still are predicated on Joyce's work. While *Dubliners 100* covers Joyce's original stories directly by revisiting, at times even subverting, the plots of the stories, it is evident that the collection pays homage to Joyce and his legacy: 'The idea was simple: fifteen contemporary Irish authors covering fifteen original stories of *Dubliners* to mark the collection's centenary.'[28] *The New Dubliners* collection, on the other hand, is more intent on representing twenty-first-century Dublin, though it too references itself as a reiteration of Joyce's *Dubliners,* 'It [*New Dubliners*] does not allow simply for a consideration of Joyce's enduring influence on Irish literature; *New Dubliners* also offers expansive, imaginative, hilarious, poignant and daring considerations of the life of Joyce's much changed capital city.'[29] This new view of the city and its denizens in this collection is highlighted by the inclusion of one of Roddy Doyle's short stories, 'Recuperation,' which he also includes in his own 2011 short story collection also set in Dublin, ***Bullfighting***, which I discuss later in this chapter.

The Commodification of Joyce

In his 2014 American Conference for Irish Studies (ACIS) plenary address, 'Disappearing Ireland,' Declan Kiberd warned of the dangers of fetishising the past and turning it into a commodity, citing *The Gathering* 2013, a year-long event that 'invited friends, loved ones and connections from around the world to celebrate 5000 gatherings held in their honour'[30] as an example of commodification of Irish identity. By looking at Dublin and Ireland primarily in Joycean terms and measuring the city and its chroniclers against *Dubliners*, we are doing this very same thing. Joyce has been turned into a commercial and intellectual commodity. This is of great concern to Doyle, particularly because the appropriation of an author by academia can intimidate the general reader and dehumanise the author:

> I think it's happened with Joyce particularly. People are utterly intimidated by the prospect of reading *Ulysses*, for example, because they feel they can't do it unless they have a doctorate … And I also think he [Joyce] wouldn't have felt that everybody needs to understand every sentence that he's written Also, once it becomes the intellectual property of a group of people, sometimes there's a toll that is also attached to the humanity of the writer that is forgotten about. It's easier to do when the writer is dead, obviously.[31]

This is how Joyce has been turned into a spectre whose human flaws have been erased. These shortcomings have been minimised to the point that he is now reduced to a phantom figure. Joyce the author and his works have succumbed to the forces of economics and canonisation. Joyce, the man, however, was very much aware of his own flaws, if only in terms of how subjectively he depicted Dublin. As Ellmann remarks, '[Joyce knew that] a mirror held up to nature will reflect the holder's consciousness as much as what is reflected. He [Joyce] could quote with approval (May 16, 1907) Pater's remark, "Art is life seen through a temperament".'[32] Joyce also had a very specific goal, which Ellmann points out, 'He wished to give his contemporaries, especially his Irish ones, a good look at themselves in his polished looking-glass—as he said—but not to destroy them. They must know themselves in order to become freer and more alive.'[33]

Though he regarded his vision of Dublin as a unique one, Joyce acknowledged that he was not the only one who could write about the citizens of Dublin: 'Of *Dubliners*, he said to his brother, "The stories seem to be indisputably well done, but, after all, perhaps many people could do them as well".'[34] In fact, someone else had. This is the other secret hidden by the spectral Joyce on the ten-pound note. He was not the first writer to feature Dublin in his short stories. George Moore, like Joyce, was eager to write about his city. In the preface to his collection, *The Untilled Field*, Moore comments that he wrote it with 'the hope of furnishing the young Irish of the future with models'[35] and 'a desire to paint the portrait of my country … [and] the villages around Dublin.'[36] As Richard Robinson remarks, Joyce disdained Moore's collection, perhaps because it was published in 1903 (before *Dubliners* was written), and the two works are very similar in structure, theme, and tone: 'Moore plays the role of a pioneer in Irish prose fiction who set up a template of genre—the unified volume of short stories about Ireland—which Joyce then immediately exploited.' As one critic put it, '*Dubliners* is the second chapter of [Ireland's] moral history. The *Untilled Field* is the first.'[37] Robinson also refers to comparative studies of *The Untilled Field* and *Dubliners* and cites Heather Ingram's comment that 'through [Moore's] remarkable intertwining of the local and the foreign, the modern short story was born in Ireland.' The consensus is that 'it is almost as if Moore's volume makes a textual epigone out of the proto-modernist *Dubliners*.'[38]

Roddy Doyle's Dublin

There is a third chapter of Ireland's moral history, one that chronicles twentieth- to twenty-first-century Dublin without measuring itself against *Dubliners* or Joyce. Writing a century later, Doyle also focuses on his native city, but not because he is trying to emulate Joyce. As he states, 'my characters are out on the streets of Dublin, or in their houses in Dublin, or in a pub in Dublin. I don't do that because Joyce did that. It's how city life works.'[39] Acknowledging Doyle's talent, Toronto's *The Globe and Mail* makes the inevitable comparison to Joyce by describing Doyle's *Bullfighting* as 'probably the finest collection of Irish short stories since James Joyce's *Dubliners*.'[40] When asked if he feels that his short story collections *The Deportees* (2007) and *Bullfighting* (2011) are modern versions of *Dubliners*,

Doyle is quick to disagree: 'No, I don't ... You see, the problem is it's almost as if Joyce invented Dublin and everybody has to then be judged against Joyce and, of course, he didn't.'[41]

Although Doyle refuses to be compared to Joyce, in homage to the centenary of *Dubliners*, *The New York Times* felt it was important to pose the question 'Who are James Joyce's Modern Heirs?' to two writer-critics, as part of its new Sunday Book Review feature, Bookends. One of the respondents, Rivka Galchen, states in her first sentence that 'Joyce's work is so canonical that in some sense we are all inescapably his heirs' and in the next few lines that 'we are his heirs because we live in the spaces he built,'[42] thus claiming Joyce's absolute literary dominance. The second critic-writer to answer the question was Pankaj Mishra, who after quoting Stephen Dedalus' pronouncement to an Irish nationalist: 'You talk to me of nationality, language, religion [...] I shall try to fly by those nets,' finds a way to mention Mahatma Gandhi and Joyce in the same sentence, thus giving Joyce, the man who refused to involve himself in any of the meaningful political events of his country, the same standing as one of the greatest political activists of all times. Amit Majmudar, in the *Kenyon Review* blog, answers the same question by listing Joycean traits that would help identify his heir, some of which could easily be attributed to Doyle: 'contemporary highly local vernacular,' 'an ability to write men [the Rabbitte men] and women [Paula Spenser],' 'a wish to transcend mere nationalism and locality,' and particularly, 'to write not Britons, not Irishmen, but Dubliners.'[43] Majmudar concludes that, 'few dead writers and living writers ... demonstrate all.'[44] The cult of Joyce is alive and well and that Joyce's presence/absence still haunts many today. But not Doyle. Doyle's version of Dublin is that of the insider and successful writer who has lived there all his life and who walks its streets daily.

Unlike Joyce, Doyle is proud to be a Dubliner and has never felt the need to emigrate:

> I wondered you know, out of wanderlust, if I would have liked to move around the world a bit, but I decided, no, I was much happier working in the school so I never had to emigrate, you know? There was a point where I was the only one of my siblings, there are four of us, who lived in Ireland, and that wasn't unusual. But it wasn't an issue for me. Never. And I didn't stay because I wanted to write about Ireland because at that point I wasn't writing, you know? It just wasn't a necessity.[45]

On the other hand, Joyce made his loathing for the city clear, in a letter to his wife Nora, 'How sick, sick, sick I am of Dublin! It is the city of failure, of rancour and of unhappiness. I long to be out of it.' [46] Though his dislike of Dublin seems to make little difference now, as for many Dublin is still synonymous with Joyce. It did make a difference during his time, however. Herbert A. Kenny explains that after an American judge lifted the ban (on the grounds of obscenity) on *Ulysses* in the US, 'Joyce was a cult, the object of literary adulation on both sides of the Atlantic.'[47] Perhaps, Joyce's absence (during and after his lifetime) is a great part of what draws his literary followers. Their quest for an imaginary Ireland is an idealistic one. Ironically, while they look to the past to follow the author's trace, Joyce's own search for Ireland lay in the future. As Ellmann points out, 'it was not Ireland as it had been that attracted him, but Ireland as it might be.'[48]

Kiberd warns us against fetishising the past and turning it into a commodity.[49] By looking at Dublin primarily through the Joycean lens and measuring it and its chroniclers against *Dubliners*, we are in danger of doing the very thing that Joyce was condemning: paralysing the city. We are also turning Joyce into a commodity, both commercial and intellectual. As Doyle notes:

> You know what is the third generation of scholarship? Now people reporting, writing about the scholars. And it's like boys pissing in the snow to see who can piss farther. So I think Joyce is probably beyond saving at this stage. If you decided, if you pulled the plug on the Joyce industry, how many academics would be out of work? There would be quite a lot.[50]

For Doyle, staying in or leaving Dublin is also not necessarily a political statement for Irish writers. He recognises the complexity of why Irish authors are moved to emigrate and refuses to generalise. Instead, he offers his own experience:

> I stayed in Dublin because I liked it and then I began to write. But then when my family arrived, you know, you can't uproot children on a whim because you think, *Oh now I'll write from a different perspective. I get angry with the cultural milieu of Dublin and decide that I'll uproot myself.* Because I wouldn't just be uprooting myself, I'd be uprooting four other people and that seems very selfish. They're Dubliners, you know, my children, they're in their twenties now, two of them. So it wasn't really an option, you know? It didn't occur to me.[51]

Evidently, Doyle's roots in Dublin are deep. Both of his parents are Dubliners and he respects his children's right to grow up in the city of their birth.

Like Joyce, Doyle focuses on the urban experience and puts his native Dublin at the centre of his creative *oeuvre*. Though Doyle offers a very different vision of Dublin, a global perspective rooted in a humanistic approach to the question of belonging. His focus on Dublin and its shifts in demographics and economic life makes his works historical documents, which chronicle Dublin's transformation in the late twentieth century and beyond, particularly during the Celtic and post-Celtic Tiger eras. Doyle writes about life in his contemporary Dublin, much like Joyce did in his own time. In particular, Doyle's short story collections, *The Deportees* and *Bullfighting*, can be compared to Joyce's *Dubliners*, in that they constitute time capsules of Dublin during their respective authors' lives. Doyle chronicles Dublin's coming to terms with the profound changes brought by inward migration, specifically the radical transformation of traditional white and Catholic Irish cultural and national identity into a multicultural and global one.

Doyle's work is the question of identity as it is formed through circumstances, relationships, and encounters. This identity is constantly shifting and questioning its traditional constants of nationality and race. Doyle explores this question in his short story collection *The Deportees* by chronicling the cultural intersections and tensions between the native Irish and the new immigrants. These stories promote 'a shift from 'multiculturality' (the disconnected and passive coexistence with ethnic minorities) to 'interculturalism' (the active exchange between different cultures and ethnicities).'[52] The Dublin that Doyle writes about is an Ireland that Joyce would not recognise. Doyle chronicles the city's transformation and identity migration towards a hybrid identity that encompasses a new vision of immigration with Ireland in the role of host country.

Doyle's own contribution to the acceptance of this new Irish hyphenated identity is exemplified by his on-going contributions to immigrant magazine *Metro Éireann* and his collection of eight short stories, *The Deportees*, initially written for this magazine. The stories trace the acculturation of African and Eastern Europe immigrants to Ireland and the inevitable encounters between the newcomers and the Irish natives: 'Today, one in every ten people living in Ireland meets someone who has come to live here … Today, one in every ten people living in Ireland wasn't born here. The story-someone new meets someone old-has become an unavoidable one.'[53] For many Irish, this has felt a bit like a Trojan invasion: 'It happened, I think, some time in the mid-90s. I went to bed in one country and woke up in a different one … It took getting used to.'[54]

Doyle's goal was to create a bridge between the old and the new Irish by launching a series of short stories that, as Doyle readily admits, have been written as propaganda:

> With *The Deportees*, I was very keen that the new to the country would rub shoulders with the people here and vice versa. Inevitably, that means people meeting. I think my stories are probably less grim, a little bit more hopeful perhaps than those of *Dubliners*. There's a propaganda element to *The Deportees* stories, insofar as I'm saying, "Isn't this great?" It may well be, I am not sure. I haven't read those stories since I wrote them, but I don't know how well they would stand up to other work that I've done because of that propaganda element—which is very, very deliberate. When you're delivering 800 words a month, there's only room for so much, and again there was the propaganda part.[55]

In that, just like Joyce, Doyle has his own agenda and makes Dublin in his own image, though the time capsules that the two authors chronicle are very different. While Joyce focuses on the paralysis that envelops Dublin in *Dubliners* and Doyle the new vitality that the inward migration brings to the city in *The Deportees*, they are both motivated by their circumstances—Joyce unhappy in a city that does not appreciate his talent and Doyle content in a new Dublin that welcomes his writing. However, the optimism of Doyle's stories did not translate to his real-life intercultural relationship with Irish Nigerian co-writer Bisi Adigun who sued Doyle and the Abbey Theatre for alterations to their adaptation of J.M. Synge's *The Playboy of the Western World*,[56] clearly questioning Doyle's optimism for interracial harmony and demonstrating that Dublin is still coming to terms with its new intercultural identity.

Doyles' collection *Bullfighting* has a narrower focus and grimmer undertone than *The Deportees*, though the stories still manage to find hope in the darkness. Just as Doyle mentioned waking up in a new Ireland in the foreword of *The Deportees*, so do his characters in *Bullfighting* wake up in a different land. Rather than finding themselves in a multicultural, young, and vibrant Ireland experiencing a new economic boom, the characters in *Bullfighting* find themselves in middle-aged bodies, surrounded by older contemporaries and adult children, in an Ireland that is now in full recession. Here, Doyle no longer focuses on the plight of African and Eastern Europeans' identity in Ireland as he did in *The Deportees*, but instead investigates the male Irish psyche during the midlife crisis, which reflects the country's economic crisis. Typical of many of Doyle's characters, the men take a stoical humorous view of their predicament.

Thus, in *Bullfighting*, Doyle's gaze shifts from the external metamorphosis of the country's new national, cultural, and racial Irish identity of *The Deportees* to the internal transformation of the Irish middle-aged male psyche. Doyle captures his protagonists in the middle of their struggle to come to terms with their new and older identity. As a result, the personal supersedes the political, and rather than exploring new vistas through diverse cultural encounters, in *Bullfighting*, the characters encounter themselves, or rather a different version of themselves.

The stories explore new territory in terms of resilience and strength. In the face of personal entropy and economic recession, Doyle's characters rise to the challenge of migrating into a new identity that is just as foreign to them as the new African and Eastern European immigrants were to their Irish counterparts in *The Deportees*. In this, these *Bullfighting* stories, although firmly set in the psyche of the modern Irish male, are universal and perhaps do more to bridge differences between race and nationality than those of *The Deportees* as they chronicle the inevitable and quietly epic male journey from youth to middle age.

One example is the protagonist of 'The Slave,' who, confronted with a rat in his kitchen, uses the encounter as an opportunity to fight his demons by equating the rat with middle age and the intrusion of nature or mortality in his tranquil home life. Though fear first paralyses him and taunts him, he emerges victorious and defiant:

> I recognize what is going on in my head, what's been going on for a while, actually, on and off. It's middle age. I know that. It's getting older, slower, tired, bored, useless. It's death becoming something real … But you can still hang on. And I'd been doing all right … And so it is. Only, it has to be protected. If you find a rat in your kitchen the world stops being a straight-forward, decent place for a while. You have to take it back. And that's what I'm doing. Taking it back. And I'm getting there … Fuck the rat…[57]

The narrator is ready to take back his home and himself, in that he will not let the intruder that is middle age rob him of his life and identity. The first and last stories of the *Bullfighting* collection, 'Recuperation' and 'Sleep,' also offer hope. In 'Recuperation,' a sick man is on his way to health, a positive way to begin the collection. This is very different from Joyce's *Dubliners*, which begins and ends with stories that focus on death and paralysis. The first story 'Sisters' sets the dark tone for the collection, announcing the hopelessness and paralysis of Joyce's Dublin through the epiphany that the death of a priest brings a schoolboy:

> There was no hope for him this time: it was the third stroke … He had often said to me: *I am not long for this world*, and I had thought his words idle. Now I knew they were true. Every night as I gazed up at the window I said softly to myself the word *paralysis*. It had always sounded strangely in my ears … But now it sounded to me like the name of some maleficent and sinful being. It filled me with fear, and yet I longed to be nearer to it and to look upon its deadly work.[58]

The schoolboy is fascinated by the state of paralysis that surrounds him to the point that he cannot shake its attraction, even though he is fully aware of its danger. In this story, Joyce lays out the climate of the Dublin he leaves behind and chronicles the hold of the old religious traditions that fetter its inhabitants.

The last story of *Dubliners* also features death and paralysis, though it uses snow imagery to illustrate Gabriel's epiphany:

> It [snow] was falling on every part of the dark central plain, on the treeless hills, falling softly upon the Bog of Allen and, farther westward, softly falling into the dark mutinous Shannon waves. It was falling, too, upon every part of the lonely churchyard on the hill where Michael Furey lay buried. It lay thickly drifted on the crooked crosses and headstones, on the spears of the little gate, on the barren thorns. His soul swooned slowly as he heard the snow falling faintly through the universe and faintly falling, like the descent of their last end, upon all the living and the dead.[59]

At the end of the story, Gabriel finds himself between the world of the dead and the living with a vision of Ireland's paralysis represented by the snow that envelops everything, from the macrocosm of the different landscapes of the island to the microcosm of the individual graveyard where the man his wife once loved is buried. Nothing escapes the cold snow that freezes in place Joyce's Ireland.

In another *Bullfighting* story, 'Sleep,' a father who has just found out he has cancer of the colon, but hasn't told his wife yet, remains optimistic in the aftermath of his child's frightening asthma attack. Tom watches his wife and child sleep as he courageously predicts a good life for himself and his son Aaron, describing the rites of passage of his boy's life as he grows into a man, armed with his inhaler:

> Aaron would be fine. He'd get into fights, he'd play his football. He'd go drinking in St. Anne's Park, in the pissing rain, with his inhaler in his pocket. He'd join a band, he'd smoke, he'd stroll up Kilimanjaro. He'd come home one morning and tell them they were going to be grandparents, and make them both shockingly happy.[60]

By envisioning an ordinary future for his sick child, the father succeeds in reclaiming his child's life as well as his own. He will not let disease take away his hope and future.

A future is just what the eponymous character of Joyce's story 'Eveline' cannot procure for herself. Faced with the opportunity to leave behind her life taking care of an abusive father to get married to a kindly man, Frank, and start a new life in Buenos Ayres, she freezes:

> She felt her cheek grow cold out of a maze of distress, she prayed to God to direct her, to show her what was her duty….
>
> No! No! No! It was impossible. Her hands clutched the iron in frenzy. Amid the seas she sent a cry of anguish!
>
> …
>
> She set her white face to him [Frank], passive, like a helpless animal. Her eyes gave him no sign of love or farewell or recognition.[61]

Eveline is incapable of moving forward, trapped by religion. Her fear strips her of her humanity to the point that she is incapable of any type of action, mummified in Joyce's paralysing Dublin.

Doyle too explores the Irish psyche, but through its displacement during a time of radical changes (both internal and external) with his own brand of optimism. He re-invents Irish identity as an on-going process of grafting and hybridisation. Doyle maps a new geography, a personal innerscape that takes precedence to all socio-political and economic issues of modern Ireland through a series of small epiphanies.

Doyle's collections *The Deportees* and *Bullfighting* stand on their own as time capsules of life in the now glocal, multicultural Dublin of the twenty-first century that encompasses both the local and the global identity that now defines this city and reminds us that Dublin does not belong to Joyce or anyone in particular. The fact that Doyle is writing in a new global context questions Joyce's hold on Dublin and opens new visions of the city beyond *Dubliners*. Not that Joyce's Dublin is to be erased. Rather, it needs to be seen as what it is: a chapter of the history of the city written by a self-exile discontented by the lack of opportunities for Irish writers in an Ireland dominated by the church and the British ascendancy. In the new post-Celtic Tiger Ireland, Doyle has the luxury to remain in Ireland and publish his works in a much more welcoming climate. This explains the hopeful tone of his works and the freedom in his subject matter, primarily the north Dublin working class. By writing about Dublin on his own terms, Doyle liberates the city and its writers from the tyranny of the author who wilfully abandoned it, but still controls it beyond space and time, much like an absentee landlord who is still collecting his rent, or better yet, a ghost that still haunts its streets and minds.

Notes

1. Andrew Kincaid 'Down These Mean Streets': The City and Critique in Contemporary Irish Noir *Éire-Ireland* 45, no. 1 (2010): 44.
2. Andrew Kincaid 'Memory and the City: Urban Renewal and Literary Memoirs in Contemporary Dublin' *College Literature* 32, no. 2 (2005): 39.
3. Margaret Hallissy, *Understanding Contemporary Irish Fiction and Drama* (Columbia, South Carolina: University of South Carolina Press, 2016): 8.
4. Jacques Derrida. *Specters of Marx: The State of the Debt, the Work of Mourning and the New International* (New York and London 1994): xix.
5. Petra Eckhard 'Chronotopes of the Uncanny: Time and Space in Postmodern New York Novels, Paul *Auster's City of Glass* and Toni Morrison's *Jazz.*' *European journal of American studies* [Online], Reviews 2012-1, document 8 (12).
6. Colin Davis "État Present: Hauntology, Spectres and Phantoms." *French Studies* 59, no. 3 (July & Aug. 2005): 373.
7. Derrida *Specters of Marx*, 20.
8. Ibid., 20–21.
9. Richard Ellmann, *Four Dubliners: Wilde, Yeats, Joyce, and Beckett* (New York: G. Braziller, 1987), 68.
10. Michael Malouf, 'Forging the Nation: James Joyce and the Celtic Tiger.' https://english.chass.ncsu.edu/jouvert/v4i1/malouf.htm.
11. Luke Gibbons, *Joyce's Ghosts: Ireland, Modernism, and Memory* (Chicago: University of Chicago Press, 2015), 20.
12. BBC News, 'Dublin's Floozy in the Jacuzzi returns to the city.' http://www.bbc.com/news/uk-northern-ireland-12579655.
13. Qtd in Davis 373–374.
14. Michael Malouf, 'Forging the Nation: James Joyce and the Celtic Tiger.' https://english.chass.ncsu.edu/jouvert/v4i1/malouf.htm.
15. Ibid.
16. Davis, 'État Présent', 377.

17. Qtd in Davis 374.
18. Jeri Johnson, 'Literary Geography: Joyce, Woolf and the City.' *City.* 45, no. 2 (2000): 199.
19. Eva Roa White, 'From Emigration to Immigration: Irishness in *The Irish Short Stories of Sarah Orne Jewett* and Roddy Doyle's *The Deportees.*' *In the Wake of the Tiger: Irish Studies in the Twentieth-First Century.* Eds. David Clark and Rubén Jarazo Álvarez. Irish Studies Series. Oleiros (La Coruña): Netbiblo, 2010. 104.
20. Ibid., 104.
21. Herbert A. Kenny, *Literary Dublin: A History* (Dublin: Gill and Macmillan, 1991), 168.
22. Luke Gibbons, *Joyce's Ghosts: Ireland, Modernism, and Memory* (Chicago: University of Chicago Press, 2015), 11.
23. Ibid.
24. Ibid., 11–12.
25. Kenny, *Literary Dublin*, 172.
26. Rubén Jarazo Álvarez, 'Managing Culture in Ireland: Literary Tourism and James Joyce.' In *New Perspectives on James Joyce: Ignatius Loyola, Make Haste to Help Me* (Bilbao, Spain: Universidad de Deusto, 2009), 238–239.
27. Ellmann, *Four Dubliners*, 66.
28. Thomas Morris, *Dubliners 100* (Dublin: Tramp Press, 2014), viii.
29. Oona Frawley, *New Dubliners* (New York: Pegasus Books), vi.
30. Declan Kiberd, Plenary address. 'Disappearing Ireland,' 2014 ACIS/CAIS conference UCD–Dublin.
31. Eva Roa White, "Roddy Doyle's Dublin: An Interview." *Breac*, August 1, 2016. https://breac.nd.edu/articles/roddy-doyles-dublin/.
32. Ellmann, *Four Dubliners*, 69.
33. Ibid., 88.
34. Ibid., 80.
35. George Moore, *The Untilled Field* (London: T. Fisher Unwin, 1903), xvii.
36. Ibid., xviii.
37. Richard Robinson. *'"That Dubious Enterprise, the Irish Short Story": The Untilled Field and Dubliners.' James Joyce in the Nineteenth Century.* Ed. John Nash (Cambridge: Cambridge University Press, 2013), 46.
38. Ibid.
39. Eva Roa White, 'Roddy Doyle's Dublin.' https://breac.nd.edu/articles/roddy-doyles-dublin/.
40. John Doyle. "Bullfighting by Roddy Doyle' Toronto *Globe and Mail*, April 15, 2011. https://www.theglobeandmail.com/arts/books-and-media/bullfighting-by-roddy-doyle/article576543/.

41. Eva Roa White, 'Roddy Doyle's Dublin.' https://breac.nd.edu/articles/roddy-doyles-dublin/.
42. Rivka Galchen and Pankaj Mishra, 'Who Are James Joyce's Modern Heirs? February 2, 2014. https://www.nytimes.com/2014/02/02/books/review/who-are-james-joyces-modern-heirs.html (accessed June 14, 2016).
43. Amit Majmudar, *Kenyon Review* blog January 29, 2014) http://www.kenyonreview.org/conversation/amit-majmudar/ (accessed June 14, 2016).
44. Ibid.
45. Eva Roa White, 'Roddy Doyle's Dublin.' https://breac.nd.edu/articles/roddy-doyles-dublin/.
46. Letters, 22 August 1909.
47. Kenny, *Literary Dublin*, 172.
48. Ellmann, *Four Dubliners*, 78.
49. Declan Kiberd, 'Disappearing Ireland.'
50. Eva Roa White, 'Roddy Doyle's Dublin.'
51. Ibid.
52. Eva Roa White 'Who's Irish?: Roddy Doyle's Hyphenated Identities.' In *Literary Visions of Multicultural Ireland: The Immigrant in Contemporary Irish Literature, 95–107.* Ed. Pilar Villar-Argáiz (Manchester: Manchester University Press, 2014), 96.
53. Roddy Doyle. *The Deportees* (Johnathan Cape: London, 2007), xiii.
54. Ibid., xi.
55. Eva Roa White, 'Roddy Doyle's Dublin.' https://breac.nd.edu/articles/roddy-doyles-dublin/.
56. Ronan McGreevy. 'Abbey 'to pay €600,000' in dispute over play copyright.' https://www.irishtimes.com/news/abbey-to-pay-600-000-in-dispute-over-play-copyright-1.1255531.
57. Roddy Doyle. *Bullfighting* (New York, NY: Viking Penguin, 2011), 65–66.
58. Joyce, James. *Dubliners* Centennial edition (New York, NY: Penguin, 2014), 1.
59. Ibid., 194.
60. Ibid., 214.
61. Ibid., 29.

References

'Dublin's Floozy in the Jacuzzi returns to the city.' *BBC News.* February 25, 2011. http://www.bbc.com/news/uk-northern-ireland-12579655. Accessed June 14, 2016.

Davis, Colin. 'État Present: Hauntology, Spectres and Phantoms.' *French Studies* 59, no. 3 (July & August 2005): 373–379.

Derrida, Jacques. *Specters of Marx: The State of the Debt, the Work of Mourning and the New International.* New York and London, 1994. http://cnqzu.com/library/Economics/marxianeconomics/DerridaJacques-SpectersofMarx.pdf.

Doyle, Roddy. *The Deportees.* London: Johnathan Cape, 2007.

———. *Bullfighting.* New York, NY: Viking Penguin, 2011.

Eckhard, Petra. 'Chronotopes of the Uncanny: Time and Space in Postmodern New York Novels, Paul *Auster's City of Glass* and Toni Morrison's *Jazz.*' *European Journal of American Studies* [Online], 2012: 205. http://ejas.revues.org/9706. Accessed June 14, 2016.

Ellmann, Richard. *Four Dubliners: Wilde, Yeats, Joyce, and Beckett.* New York: G. Braziller, 1987.

Frawley, Oona, ed. *New Dubliners.* New York: Pegasus Books, 2005.

Galchen, Rivka, and Pankaj Mishra. 'Who Are James Joyce's Modern Heirs?' February 2, 2014. https://www.nytimes.com/2014/02/02/books/review/who-are-james-joyces-modern-heirs.html. Accessed June 14, 2016.

Gibbons, Luke. *Joyce's Ghosts: Ireland, Modernism, and Memory.* Chicago: University of Chicago Press, 2015.

Hallissy, Margaret. *Understanding Contemporary Irish Fiction and Drama.* Columbia, SC: University of South Carolina Press, 2016.

Jarazo Álvarez, Rubén. 'Managing Culture in Ireland: Literary Tourism and James Joyce.' In *New Perspectives on James Joyce: Ignatius Loyola, Make Haste to Help Me!*, ed. Ma Luz Suárez Castiñeira (ed. and preface), Asier Altuna García de Salazar (ed. and preface), and Olga Fernández Vicente (ed. and preface), 233–239. Bilbao: Universidad de Deusto, 2009.

Johnson, Jeri. 'Literary Geography: Joyce, Woolf and the City.' *City* 4, no. 2 (July 2000): 199–214. https://doi.org/10.1080/13604810050147820.

Joyce, James. Letters, 22 August 1909.

———. *Dubliners.* Centennial ed. New York, NY: Penguin, 2014.

Kenny, Herbert A. *Literary Dublin: A History.* Dublin: Gill and Macmillan, 1991.

Kiberd, Declan. Plenary address. 'Disappearing Ireland.' 2014 ACIS/CAIS conference UCD-Dublin.

Kincaid, Andrew. 'Memory and the City: Urban Renewal and Literary Memoirs in Contemporary Dublin.' *College Literature* 32, no. 2 (2005): 16–42.

———. 'Down These Mean Streets': The City and Critique in Contemporary Irish Noir.' *Éire-Ireland* 45, no. 1 (2010): 39–55.

Majmudar, Amit. *Kenyon Review.* Blog. January 29, 2014. http://www.kenyonreview.org/conversation/amit-majmudar/.

Malouf, Michael. 'Forging the Nation: James Joyce and the Celtic Tiger.' *Jouvert: A Journal of Postcolonial Studies* 4, no. 1 (1999). https://english.chass.ncsu.edu/jouvert/v4i1/malouf.htm. Accessed June 16, 2017.

McGreevy, Ronan. 'Abbey "to Pay €600,000" in Dispute Over Play Copyright.' *The Irish Times* [Online], 2013. https://www.irishtimes.com/news/abbey-to-pay-600-000-in-dispute-over-play-copyright-1.1255531. Accessed January 1, 2018.

Moore, George. *The Untilled Field*. London: T. Fisher Unwin, 1903.
Morris, Thomas. Ed. *Dubliners 100*. Dublin: Tramp Press, 2014.
Robinson, Richard. 'That Dubious Enterprise, the Irish Short Story: The Untilled Field and Dubliners.' In *James Joyce in the Nineteenth Century*, ed. John Nash, 46–60. Cambridge: Cambridge University Press, 2013.
White, Eva Roa. 'From Emigration to Immigration: Irishness in *The Irish Short Stories of Sarah Orne Jewett* and Roddy Doyle's *The Deportees*.' In *The Wake of the Tiger: Irish Studies in the Twentieth-First Century*, ed. David Clark and Rubén Jarazo Álvarez, 103–110. La Coruña: Netbiblo, 2010.
———. 'Who's Irish?: Roddy Doyle's Hyphenated Identities.' In *Literary Visions of Multicultural Ireland: The Immigrant in Contemporary Irish Literature*, ed. Pilar Villar-Argáiz, 95–107. Manchester: Manchester University Press, 2014.
———. 'Roddy Doyle's Dublin: An Interview.' *Breac* [Online], 2016. https://breac.nd.edu/articles/roddy-doyles-dublin/. Accessed May 24, 2017.
Woolf, Virginia. 'Literary Geography.' In *The Essays of Virginia Woolf*, ed. Andrew McNeillie, 35. London: Hogarth Press, 1986.

CHAPTER 3

That Limerick Lady: Exploring the Relationship Between Kate O'Brien and Her City

Margaret O'Neill

In 1926, following the success of Kate O'Brien's first play, *Distinguished Villa*, in London, *The Limerick Leader* boasted:

> Limerick is renowned for the splendour of its girls, and Miss O'Brien, with her dark beauty, is no exception to the Shannon boast. She comes from one of those Irish families of bountiful numbers that win distinction in many fields—her men-folk are open-air men in that invincibly sporting county, and some of the clan, as O'Brien's should be, are redoubtable politicians. After a course at the National University she lived much abroad, teaching the foreigner, as an Irishwoman should, how to speak English: she worked in Manchester, in the best of all schools, at translation [...] and now is secretary of the Sunlight League. As she is only twenty-seven, the future is at her feet.[1]

The tension between romanticised images of the Irish colleen, represented here, and the woman artist exploring the world beyond social and sexual expectations, is represented throughout O'Brien's work. Born and raised in Limerick and then a boarder at Laurel Hill Convent school in the city,

M. O'Neill (✉)
University of Limerick, Limerick, Ireland
e-mail: margaret.oneill@ul.ie

M. Beville, D. Flynn (eds.), *Irish Urban Fictions*, Literary Urban Studies, https://doi.org/10.1007/978-3-319-98322-6_3

much of O'Brien's writing reflects upon the city, particularly the development of its social milieu over the late nineteenth and early twentieth centuries, as well as the expansion and commercialisation of the 1960s. Reminisces of Limerick, thinly disguised as the fictional Mellick, are found in all but two of her nine novels, as well as in O'Brien's journalism and autobiographical writing. Though she travelled and explored settings familiar in a European paradigm of modernism, Limerick remained a central concern and the influence of the city may be read across her life and works. The ambivalent nature of this relationship is illuminated in O'Brien's fictions; as her characters journey in the pursuit of personal desire, beyond the context of family, religion, and society, they are ever grounded by a sense of place, both psychologically and geographically. In her travelogue *My Ireland*, O'Brien uses the term *locus standi* to acknowledge how Limerick influences her perspective on the world. She quotes Spanish philosopher George Santayana in saying, 'the freest spirit must have some birthplace, some *locus standi* from which to view the world and some innate passion by which to judge it. Modestly I say the same for my relationship with Limerick. It was there that I began to view the world and to develop the same passion by which to judge it.'[2] In this passage, O'Brien understands her perspective on the world as induced by multiple positions—Catholic bourgeois, family structure, education, geography, history. This multivalenced positionality permits O'Brien to expand. This chapter will explore the complex issue of the relationship between O'Brien and her home city, exploring the social changes and tensions of modernity, the creation of the Catholic middle class, and the position of the woman and artist within it. It will argue that, neither entirely belonging nor fully exiled from home, O'Brien retains Limerick as a centre point while also looking away from it to other European cities on the world map of her fiction. In this manner, O'Brien may be positioned within a distinct mode of Irish cosmopolitanism identified by Nels Pearson, which is characterised by the 'need (or ability) to interlace one's still-forming Irish identity with one's often capacious global perspective.' Defying absolutes of dislocation, this 'doubly transformative' orientation resists 'the closure of Irish backgrounds even as it looks beyond them.'[3] Reading O'Brien in this light, this chapter will consider how her writings contribute to debates about the spatial and temporal boundaries of modernism, particularly the role of 'peripheral' cities in the formation of a supposedly metropolitan modernist culture. O'Brien's emphasis on the regional as well as the transnational draws attention to multiple modernities and provides for expanding our understanding of Irish and women's modernisms and the complexities of the modern.

O'Brien's earliest work as a professional writer is the play *Distinguished Villa*. Its themes as well as the author's apprehension surrounding its reception present telling insights into O'Brien's relationship with her home city. The play was first staged as a one-off performance on 2 May 1926. It attracted the attention of producers seeking a new West End play and two months later, on 12 July, opened in the Little Theatre. *Distinguished Villa* had a successful run alongside Sean O'Casey's *The Plough and the Stars*, which was playing in the New Theatre. Indeed, O'Casey sent a telegram saying 'Dublin ventures to salute Limerick.'[4] Although O'Brien would find her niche as a novelist, the radical themes presented in this drama would progress in the fiction to follow, namely, social appearances versus private reality, an ambivalent relationship with home, and personal desire in the context of family obligations. This emphasis on the intimate and the local would flow into O'Brien's later work, coming together with transnational associations to kindle her imagining of different social spaces. It is striking that none of the love relationships in *Distinguished Villa* are destined to work out and, far from 'distinguished,' they spiral into relationships that are desperate and self-serving. In this play, Natty and Mabel have been married for 11 years; however, their relationship is one of little affection. Frances, who lives in the house as a lodger, loves John. However, John must marry Mabel's younger sister, who is pregnant with a child that he believes wrongly to be his. Natty, desperate for kindness and human affection, commits suicide at the end of the play.

Though set in a London suburb, considering that O'Brien's later work explores the Irish Catholic middle class, this early work might be read as a comment on the author's social background in Limerick. Indeed, Eibhear Walshe has noted that the character of Frances, as O'Brien describes her, is very close to a self-portrait: 'She looks about twenty-eight, and is slim and tall. Her dark brown hair is closely shingled. Her face is leanly cut in noble boyish lines […]. She carries one or two books and writing materials and seems a little diffident.'[5] Later works of O'Brien's would turn to Limerick more overtly, inspired by lessons learned from the art of theatre. As Anthony Roche notes, 'To the extent that the Irish theatre had admitted Ibsen, it had Celticised him by taking the character Nora out of a contemporary drawing-room and removing her to a lonely peasant cottage. Kate O'Brien restores Ibsen's original setting, the claustrophobic living-quarters of the upper middle-class bourgeoisie, and with it much of his social critique.'[6] Ibsen's *A Doll's House*, as Toril Moi describes, 'calls for a radical transformation [forvandling], not just, or not even primarily, of laws and institutions, but of human beings and their ideas of love.'[7] Similarly criticising institutionalist ideals about human nature and society, and engaging with British and European cultures, O'Brien's modernism belies universalising ideas about Irish modernism, which has historically been determined by a perception of Ireland as restricted by imperial governance, traditionalism, and geographic and political isolation.

Limerick city is the birthplace of this important Irish writer, born in 1897 to a Catholic bourgeois family. O'Brien was raised and educated in Limerick, retaining affection for the city and county throughout her life. Due to the rapidity of social change in the years before and after her birth, O'Brien's Limerick presented a dramatically different cultural and geographical landscape to that of her grandparents, and she would chart this development in her fiction. As Michael O'Toole describes, 'She was the only Irish writer of her generation to portray seriously the new Irish mercantile class that was both Catholic and provincial—the so-called new merchant princes who had sprung from the toughest peasants of the nineteenth century.'[8] By the end of the nineteenth century, a prosperous Catholic bourgeoisie emerged in the provincial towns. Against this backdrop, O'Brien grew up in a red-brick villa, Boru House, on Mulgrave Street in Limerick. Her father, Thomas O'Brien, ran a successful business breeding and trading horses. On his side of O'Brien's heritage lies the Ireland of the twentieth century, an emerging nation characterised by political unrest, social change, and Catholic mores. In her autobiographical work *Presentation Parlour*, O'Brien describes how he was 'a child of the post-famine evictions,' as his father was turned out of a small-holding in around 1850, whereupon he made his way as far as Limerick with his family on an ass-cart.[9] In search of a new living, he turned to the horse trade, for which he had great talent, and 'in a very short time he made himself an expert on the breeding of blood stock, hunter and harness thoroughbreds.'[10] By the time his son Tom, O'Brien's father, was leaving school, 'his father was an admired citizen of Limerick, and was building himself a brick villa to flank the paddock and stables where he had based his business of blood stock breeding.'[11] Her grandfather's legacy would be echoed in O'Brien's first novel, *Without My Cloak*, wherein the Considine family forges a path to prosperity and gentility in late nineteenth-century Mellick, O'Brien's fictional Limerick.

The other side of O'Brien's heritage, which would also inspire her fiction, is inherited from Catherine Thornhill, her mother. The Thornhills were of Cromwellian stock. In *Presentation Parlour*, O'Brien describes how a soldier of Cromwell's army in the 1650s settled and married in Limerick, with the family eventually becoming Catholic. She continues, 'three hundred years later to the very decade, my grandfather, Patrick Thornhill, then only a boy, was having to fight with guile and energy—in common with uncles, brothers, cousins—to hold on to the same lands which the original conqueror had given to their Plantation ancestor.'[12] This reversal of circumstances, from conquest to loss, may be read to inform the tone of O'Brien's fiction. As Eavan Boland explains, O'Brien, in describing her maternal inheritance so, 'fixes herself, or at least a fragment of herself, in that maverick, Cromwellian line of people—Yeats, Moore, Maria Edgeworth to name only a few [...] it enables her to partake in that Ireland of grace and gesture so mocked by Maria Edgeworth in "Castle Rackrent," so mourned by Yeats in all his lyrics.'[13] Indeed, a shadow of the cultural haunting of personal consciousness associated with the Big House novel may be read on O'Brien's literary depictions. This haunting may be observed, not in the loss of a way of life, associated with the Anglo-Irish ascendency, but in a loss of selfhood within the socially ordered life of the Catholic bourgeoisie. Such loss is illuminated, for example, in Theresa Mulqueen's regrets as she lies dying from cancer in O'Brien's second novel *The Ante-Room*. Drawing on her inheritance from her mother and her father, then, O'Brien's work speaks to the interplay of religious, political, and cultural formations as they live on in Irish society and marks the multiple temporalities, spaces, and subjectivities to which they give rise.

O'Brien's narratives are interjected with characters and events that echo the author's life and experiences. Such occasions of reminiscence, as they are re-imagined in novel form, raise questions as to the extent to which life can be consciously shaped, as opposed to socially determined. O'Brien's mother, Catherine, died from cancer when O'Brien was only five years of age. At this point, she would become the youngest boarder at the French convent school Laurel Hill in Limerick, an experience that would inform her 1941 novel *The Land of Spices*. In 1916, her father passed away, and O'Brien embarked to University College Dublin (UCD), later travelling and working in England, Europe, and the US. This period includes a formative year as a governess in Spain, reflected in her 1936 novel *Mary Lavelle*. Though she left Limerick and Laurel Hill, a social and familial environment in which men's and women's roles were rigidly defined, these social mores, characters, and institutions would influence O'Brien's novels for decades to come, as she would bring her memories to bear on her fiction. In her vivid portrayal of middle-class life, the laws of social and sexual convention are illuminated, as outward gestures are shown to belie inner passions, and the impetus to personal, political, and sexual freedom is central. Neither O'Brien's travel writing nor her fictions are autobiographical. However, the 'scenes, places, moods and manners' that she recalls in *My Ireland* are telling of the manner in which her novels, similarly, reflect on a formative experience of belonging, and yet not belonging, to the Irish city, and an impetus to share this journey of self-expression through her fiction.[14]

In the context of the development of the Irish novel and, specifically, considerations of the developing urban space, O'Brien's influence is central. Traversing the city, however, and writing within the tradition of the *bildungsroman*, the artist faces particular challenges in negotiating these male-dominated social and cultural spaces. Boland, herself a major Irish poet, has dealt specifically with issues of national identity and the position of the woman and artist in a history and culture in which the tradition of womanhood is idealised as a cultural icon—the passive object of the national story—and in which the historic vocation of the poet is largely configured as male. In her autobiographical work *Object Lessons: The Life of the Woman and the Poet in Our Time*, Boland illuminates the position of a woman artist in the literary culture of a city imbued with male assumptions, drawing on the lessons of her own artistic journey in the city, in this case Dublin. As she states, 'in the Dublin I first knew, it was possible to be a poet, permissible to be a woman and difficult to be both without flouting the damaged and incomplete permissions on which Irish poetry has been constructed.'[15] She speaks of missing stories and histories—narratives of women and poets—and a desire to read the traces of those who have been and are gone. Turning to O'Brien, then, and Limerick, it is significant that, as Boland states elsewhere, the author 'did not seek to disassemble that trope [the passive feminine] in any immediate or obvious way. But because her destiny as an imagination would be to explore a delicate compound of conscience and sexuality [...] she would be a powerful part of its unravelling.'[16] There is in O'Brien's work an emphasis upon the individual woman, on inner life, and the difficulty of achieving selfhood in the Irish city. At the same time, O'Brien's characters' lives are acted out against the backdrop of the spires, castles, and tall houses of Limerick, in addition to the traditional centres of European modernism. Her novels embrace modernity both in Ireland and elsewhere, which sets them in opposition to dominant isolating constructions of the Irish literary tradition. As Matthew Reznicek argues, 'The distinction between a backward and nostalgic Ireland and a modern England, based on ideas of Irish space and place, has dominated the definition of Irish texts.'[17] Reznicek illustrates, in his exploration of the role of the Paris cityscape, how an emphasis on Irish literature written by women in the European metropolis challenges the cultural assumptions and gendered spheres that define modernity and provides a different way of thinking about the development of modern subjectivity. Turning to Limerick, one might argue that as O'Brien travels and explores the modernity associated with elsewhere, she also casts light on the modernisation of her homeland. Her work provides for considering the aesthetics of modernity in the Irish city in a new light, challenging dominant cultural narratives which centralise men as the agents of change, and allowing for women's diverse and active relation to history, culture and society.

Though O'Brien loved Limerick, her relationship with home was guarded and therefore she wrote about her city and middle-class life obliquely. As she states in *My Ireland*: 'I find that to write descriptively and directly about Limerick or Kilkee is but impossible—for every sentence turns in on me, becomes evocative, imaginative and private—all out of place.'[18] Written from locations such as Oxford, Rome, and eventually Roundstone in Co. Galway, O'Brien's female characters travel as their creator does, and this provides the distance necessary for self-reflection and critical consideration of home. Time and again, these fictions engage with Limerick, as Mellick, through tropes of departure and return. Her canon presents a chronology of life in the city from the end of the eighteenth century through to the early decades of the twentieth century, focusing on the world of the Catholic bourgeoisie, particularly the private lives of women within this class. Her first novel, *Without My Cloak*, was published in 1931 to enormous success. A chronicle of the Considine family over three generations, the families' rise to a position of power and influence reflects the rapid expansion and development of the town as an important centre of trade and commerce over the course of the nineteenth century. In the Prologue, Anthony Considine surveys the city nestled in its surrounding landscape:

> The Vale of Honey is a wide plain of fertile pastures and deep woods, watered by many streams and ringed about by mountains. Westward the Bearnagh hills, through whose Gap of Storm the traveller had just tramped, shelter it from the Atlantic-salted wind, and at the foot of these hills a great river sweeps about the western valley, zigzagging passionately westward and southward and westward again in its search for the sea. A few miles below him on this river's banks the traveller saw the grey blur of a town. 'That must be Mellick' [...] Villages lay untidily about the plain; smoke floated from the chimneys of parked mansions and the broken thatch of cowmen's huts [...] "The Vale of Honey!" He said softly. "I've often heard 'tis a grand, rich, easy-going place".[19]

Later novels of O'Brien's would return again to Mellick as an important touchstone. Her writings were set in the progressively less distant past, and this developing urban centre provided a point from which to examine and challenge not only the Ireland of previous generations but also the rapidly changing social and political landscape in which she was writing. Through this 'aesthetic landscape' situated on the southern border between Clare and Limerick, O'Brien, as Charles Travis describes, 'was able to trace in her prose-fiction the evolution of the Irish provincial town and consequent ascendancy of its Catholic bourgeois from just prior to the Act of Union, to the apotheosis of this class during the period marked by DeValera's constitutional plebiscite of 1937.'[20] Thus, O'Brien's writing highlights this location as one among multiple times and spaces in modernising Ireland. Centring on a supposedly provincial region, her work provides for expanding the conventional boundaries and definitions of modernism, to rethink Ireland's relationship to modernity. In *Without My Cloak*, Anthony Considine forges the beginnings of a family business in trade that will thrive, adapt, and relocate as quays, streets, and squares are built to meet the needs of the expanding city. This is reflected as the business moves from the humble origins of Lady's Lane to the lively and prosperous Kilmoney Street, and then, when Kilmoney is in decline, to the New Town. Now, two generations on, 'between Hennessy's Mills and the Passionist Church, with its back windows opening on the Dock Road and the river, and on its face a look as blank and sad as any worn by its more venerable neighbours, the store managed to exude an air of ancient permanence.'[21] Anthony's grandson and namesake takes in 'Ships and cargoes,' 'wide crossing streets,' and 'the life of King's Street, where the shops were gay.'[22] In this midst of this urban scene, 'broughams and phaëtons, splashed arrogantly through the mud, bearing wives and daughters of the town to and fro between the tall brown houses at the southern end and all the fripperies and agitations of their social habit.'[23] This brief glimpse into a passing carriage, gesturing to the life of those within, speaks to the position of women in this urban setting, in particular women of O'Brien's class, and to the question of self-expression that would be explored in increasing depth in her fiction.

Identification with the city of Limerick is for O'Brien complicated by the oppositional histories represented by her mother's and father's heritage. The castle and city walls, the Catholic and Protestant church spires, evoke a city haunted by a long history of colonial violence and struggles, political and religious divisions. This historical and geographical backdrop is echoed in the worlds of O'Brien's creation, as the women at the centre seek to make sense of their position in time and place. Her representation of modernity may be read in terms of Rita Felski's argument that 'Historical periods are not homogeneous and self-contained blocks of time governed by a single unifying spirit.' Instead, 'at any given moment, there are cross-linkages and connections, patterns of mutual influence, shared perspectives within and across specific contexts.'[24] In O'Brien's work, while locatedness, or point of view, is individual and subjective, the cultural and built environment yet holds influence. For instance, the 'Passionist Church,' beside which sits the family business in the above scene from *Without My Cloak*, speaks of the traditions and religious practices which persist alongside the forces of modernity, as this structure literally and figuratively looms over public and private life. In addition to providing an economic and religious chart of the city, O'Brien is also providing a classed map. As this scene also evokes, class divisions are part of this urban environment. Although the slums and deprivation of the city are only glimpsed, the progression of the family business to gradually more affluent streets indicates that the 'gay shops' are not available to everybody. These shopping and business districts are most likely the preserve of the Catholic bourgeoisie or Protestant ascendancy classes who inhabit the 'broughams' and 'tall brown houses.' Such public and private spaces are in addition implicitly gendered according to male and female spheres, which further complicates O'Brien's representation of temporality. As Felski states, 'Women qua women [...] have a unique relationship to time outside conventional, male-centered forms.'[25] Without positioning women within a separate time, Felski upholds that individuals may understand the same moment in history in dramatically different ways.[26] Illustrating some of the ways in which gender, sexuality, and social roles may affect an individual's relationship with time and place, O'Brien's fiction complicates the conventional boundaries of historical periodisation and indeed of Irish modernism. She is of Limerick and identifies closely with the city, and yet the urban scene encapsulated in the opening pages of *Without My Cloak* speaks to the difficulties of reconciling discourses of national, religious, and gender identity with the realities of individual desire in a modernising world. The structures and divisions of this urban space gesture to a city that represents for O'Brien at once her home, a place of shelter and possibility, but also a site of paternal control. This ambiguous positioning results in a conflicting strain in O'Brien's fiction between home and elsewhere, as her characters are compelled to travel.

As she and her characters travel, this urban space of 'Mellick' remains a centre point in O'Brien's fiction. Although the focus on the urban geography of the city in *Without My Cloak* lessens in her later work, replaced by Spanish, French, and Italian cityscapes, O'Brien continues to draw on her knowledge of Limerick's society, history, and geography while refining the scope of her narratives to further consider the topics of personal desire and sexualities in alternative destinations. *Without My Cloak*, however, represents a guide to O'Brien's literary archive in terms of connecting Ireland outward via the shipping industry and the world map of her fiction. Her writings therefore illustrate the significance of the regional as well as the transnational in Irish modernity. The 'ships and cargoes' that she mentions in the early scene of this novel draw attention to the port of Limerick in particular, as this represents a threshold between her home city and the cosmopolitans of American and European ports. Limerick thrived during the Victorian era through Empire trading, and its situation on the Shannon ensured its survival through conflict and upheaval. Its position at the Atlantic gives it a specific international and recognisable identity. All port cities bear similarities: they are micro-economies; ships are built so that they can enter other ports to trade, linking port cities without linking the countries themselves; merchants and traders encircle their domains; diversity around port cities enriched an international culture. Positioning Limerick in an international geographical and economic context, therefore, O'Brien links it with other cosmopolitan locations in her fiction, such as Rome, another port city. Thus, national boundaries are moderated in her writings in favour of an understanding of identity as both local and global. O'Brien is an urban writer, in that she is guided by her city; however, she also illustrates that a young woman in early twentieth-century Ireland was alone with her desires in the world of 'Mellick.' Her work is therefore shaped through cultural interaction and exchange, as O'Brien takes her characters elsewhere in a manner that transcends national, political, and religious subjectivity. The port in *Without My Cloak* may be read as a symbol of anticipation and possibility, as an individual could sail to new worlds, discover and exchange cultural treasures in other cosmopolitan destinations, and bring these experiences back with them.

As her novels became more experimental and further pressed the boundaries of social acceptability, two of O'Brien's books were banned in Ireland on the grounds of indecency by the Censorship of Publications Board. This censorship indicates the challenge her works posed to Church and State discourse. *The Land of Spices*, her second work to be banned, recalls the author's childhood years in Laurel Hill in Limerick. It explores the interior world of the Reverend Mother, who as a young girl returned home from school early one day to find her father 'in the embrace of love' with a man.[27] Her letters home show that O'Brien was pleasantly surprised at the reception of this book in Limerick. She was especially pleased at the good reviews and the news that her novel was selling well, to the extent that she had passed on a message from her sister Nance to her publisher Heinemann to say that her hometown was eagerly awaiting more copies of the book and could they speed things along.[28] Sadly, the book was to be banned within days of this letter. However, Lorna Reynolds states that O'Brien once said to her that 'she was sure that, if the decision was left to Limerick, she would never have been banned.'[29] This statement indicates that Irish society and Limerick in particular, in O'Brien's view, was open to art and ideas that would challenge the ideology of the Church as this prevailed in the social policy of the Irish State. It also attests to the author's enduring admiration for her home city despite her increasingly strained relationship with a religiously informed, patriarchal Irish Free State under Taoiseach Éamon de Valera.

In the writer's final finished novel, *As Music and Splendour*, published in 1958, she continues to press against social and cultural conservatism, positioning a lesbian romance as central. Set in the 1880s and 1890s, it tells the stories of Clare and Rose, who are sent from Ireland to Paris and then Rome to train as opera singers. As they travel away from home, which for Rose is a village on the border of Mellick, both engage in lifestyles and romances that are at odds with the norms and values of their homeland. Clare's lesbian relationship is an affirmative experience through which she grows in self-knowledge. At the same time, her origins and recollections provide her with a particular vantage point or *locus standi* from which to evaluate the world:

'The child heard the long-ago words. Out of the wind, my love. She was on her way, involuntarily, into horrid winds, whence no-one would call her back, come in, my love. But she knew none of that. She was on a trackless road. And she did not know that she carried armour.'[30] Clare's ability to refer to home, and her internalisation of her relationship with her grandmother, provides her with valuable 'armour' with which to help negotiate the 'trackless road' of her transition between life in Ireland and that of the operatic world. This may be read as O'Brien's most radical novel, and Clare the character who, as she travels, progresses furthest in realising her true emotional self. And yet, as with O'Brien, Clare's powerful relationship to the landscape and social and cultural background of her homeland, however ambivalent, remains central to this analysis.[31] In this fashion, O'Brien's contribution to urban modernism centres on her representations of cultural interaction and generative interconnections between the intimate and local, and the metropolitan and global.

In an article entitled 'Limerick, as I Remember It,' in 1957, O'Brien compares the city of her childhood to the developing and expanding city. She states: 'Our city is not only a beautiful piece of 18th century town planning, but it is also now clean and orderly, and without undue offence to the past it has managed to play fair with the exactions of the twentieth century.'[32] Despite her sometimes strained relationship with her home city, O'Brien retained a fond and anxious attachment to it and identified in it an air of gentility. This reference to the demands of the twentieth century speaks to an ongoing conversation in O'Brien's writings, as they reflect the developing city, between the traditions of the past and the uncertainties of modernity. David Lloyd, in *Irish Times: Temporalities of Modernity*, posits 'that the temporality of modernization in nineteenth-and early twentieth-century Ireland is rifted with formations that live on as the altered shape of practices which, rendered unviable by the inroads of colonial capitalist rationalization, find new and resistant ways to persist.'[33] Indeed, O'Brien's tendency to critique some aspects of modernity is tempered by her recognition of such lasting formations. Limerick in the mid-twentieth century saw a new wave of prosperity and expansion, as transatlantic flights at Shannon Airport opened a new gateway between Europe and America. Although O'Brien embraced such advancement, at the same time she cherished many of the associations of the past, including the shape and character of the old city and its surrounding countryside. We may read this apprehension in connection with the potential loss of individuality in an expanding city. Reznicek defines one of the central concerns that dominate representations of the cityscape as 'the inability to maintain or understand values within a system of metropolitan exchange, and, most importantly, the fear of the loss of a fixed sense of self in a metropolitan and capitalist environment wherein everything is negotiable and commodified.'[34] Such concerns may be identified in O'Brien's 1962 travelogue *My Ireland*, in which she refers to the wide streets of Limerick that 'quite monotonously now have all the usual shops of all the world, with the usual tediously celebrated packaged goods. No glossy cosmopolitan need go short of his customary fads in Limerick.'[35] She continues, 'That is what is ordinary about us—that is what seems inevitable and is sad. The unavoidable mass supply-and-demand thing which has long ago made most of the cities of America unidentifiable, interchangeable.'[36] However, this diffusion of American products also represents connections that defy distance in global modernity and an avenue through which an individual may follow their desires. Furthermore, O'Brien recognises the persistence of the intimate and the local, whereby within this cosmopolitan city she can still identify 'the private town I was a child in.'[37] In her fiction, the author therefore imbues capitalist modernity with elements of Georgian Limerick, and this overlapping of temporalities serves to fuel the ethical, sexual, and social alternatives of her imagined worlds. Drawing from Limerick, furthermore, the author insinuates that there is a certain detachment in this city from the laws of government, something perhaps outside the oppressive rule of the Irish State, from which other cities might take note. As she states in her *Irish Times* column 'Long Distances,' having extolled Limerick's virtues, mainly its 'particular quality of modesty and gentle mannerliness,' she asserts that 'In an ugly world there is still the good anarchy of Ireland. If only Dublin were more aware of the importance of that quality.'[38]

Recent studies in Irish literature posit 'a plurality of modernisms,' which, as Laura Arrington asserts, 'has enabled critics of other literatures to expand modernist studies beyond high modernism into "low" forms, or popular cultures,' and 'means that the theories of Modernism can be brought into agreement with modernist authors' various concepts of self and society.'[39] Such plurality provides for fruitful consideration of O'Brien, often overlooked as a popular romance writer, in terms of her engagement with cosmopolitan modernity and the interaction of religious, political, and cultural formations in mid-century Ireland. What comes to the fore in this perspective is O'Brien's engagement with multiple temporalities, across space and gender, as well as her representations of transnational connections across the regional and the global. Viewed this way, her work serves to contest binaries such as modern versus traditional and to challenge normalised discourses about the modern. Centralising women in this context, emphasising their mobility across spatial and social borders and boundaries, O'Brien draws attention to women's diverse and changing relationships to modernity, whether from home or elsewhere. Her writings, both fictional and discursive, reveal that though she was compelled to travel and explore, home, family, and education remained an important touchstone for O'Brien. Her family heritage, geographical location and place in time, and her reflections on the developing city over the course of the nineteenth and early twentieth centuries place O'Brien in a unique though largely overlooked position in the canon of Irish literature. Boland, in 1970, states that O'Brien provides the answer to a dilemma in Irish writing of the twentieth century. The landscapes of Yeats' Sligo are recognisable, as are the streetscapes of Joyce's Dublin. There is a division, however, between these lands, which is not a division of county or even of Anglo-Irish and Gael, but an imaginative division. As Boland asserts, 'In order for one to fully interpret the Irish literary tradition of the twentieth century one must find somewhere an imaginative achievement, a social reality which suggests the transition between the Ireland of Yeats and that of Joyce. One does not have to look further than the Mellick of Kate O'Brien's early novels to find it.'[40] And yet, O'Brien's visions of political and sexual freedom, her independent nature, her gender and sexuality, the very wilfulness of her characters, have led to her marginalisation in cultural and critical discourses. It is telling that in her older years, she was not granted the freedom of Limerick city despite calls from supporters. For instance, members of An Club Leabhar proposed that a resolution calling for this honour should go forward to Limerick City Council.[41] And yet, it was withheld. To return then to the question of belonging, there is no straightforward answer, for O'Brien's novels complicate questions of gender, sexual identity, and national belonging. However, O'Brien has suggested that despite the complexities of her relationship with her city and any responsibility she may or may not feel, she belongs regardless: 'if Limerick made me and made me love her that is Limerick's funeral.'[42] Her illustrations of women and men's changing lives in light of world events; new technology; and cultural, economic, and political developments, within which they explore the fluidity of gender, the expression of sexual desire, and possibilities for personal freedom, as yet hold the potential to complicate the politics of modernist canon-making and enact more inclusive concepts of belonging in the Irish city.

Notes

1. 'T.P. Pays Tribute to Talented Limerick Lady,' Excerpt from *T.P.'s and Cassell's Weekly*, reprinted in *Limerick Leader*, August 23, 1926, 4.
2. Kate O'Brien, *My Ireland* (London: B.T. Batsford, 1962), 148.
3. Nels Pearson, *Irish Cosmopolitanism: Location and Dislocation in James Joyce, Elizabeth Bowen and Samuel Beckett* (Gainesville: University of Florida Press, 2015), 83.
4. Sean O'Casey, qtd. in Eibhear Walshe, *Kate O'Brien A Writing Life* (Dublin: Irish Academic Press, 2006), 40.
5. Kate O'Brien, qtd. in Eibhear Walshe, *Kate O'Brien: A Writing Life* (Dublin: Irish Academic Press, 2006), 39.
6. Anthony Roche, 'The Ante-Room as Drama' in *Ordinary People Dancing: Essays on Kate O'Brien*, ed. Eibhear Walshe (Cork: Cork University Press, 1993), 89.
7. Toril Moi, 'First and Foremost a Human Being': Idealism, Theatre, and Gender in *A Doll's House*,' *Modern Drama*, 49, no. 3 (Fall 2006): 256.
8. Michael O'Toole, 'Peasants to Princes,' *Management: Journal of Irish Management Institute*, December 1995: n.p.
9. Kate O'Brien, *Presentation Parlour* (London: Heinemann, 1963), 22.
10. Ibid.
11. Ibid.
12. Ibid., 15–16.
13. Eavan Boland, 'The Treasury of Kate O'Brien's Achievement,' 'Women's Press,' *The Irish Press*, December 12, 1970, 8.
14. O'Brien, *My Ireland*, 16. As she states, 'I feel a desire to assemble in some sort the scattered impressions which Ireland has made on me over the years. No autobiography, need I say?' (ibid.).
15. Eavan Boland, *Object Lessons: The Life of the Woman and the Poet in Our Time* (Manchester: Carcanet Press, 2006), xii.
16. Eavan Boland, 'Continuing the Encounter' in *Ordinary People Dancing: Essays on Kate O'Brien*, ed. Eibhear Walshe (Cork: Cork University Press, 1993), 21.
17. Matthew, L Reznicek, *The European Metropolis: Paris and Nineteenth-Century Irish Women Novelists* (Clemson University Press, 2017), 5.
18. O'Brien, *My Ireland*, 178.
19. Kate O'Brien, *Without My Cloak* (New York: Doubleday, 1931), 3–5.
20. Charles Travis, 'The 'Historical Poetics' of Kate O'Brien's Limerick: A Critical Literary Geography of Saorstát Éireann and the 1937 Bunreacht na hÉireann Plebiscite,' *Irish Geography*, 42, no. 3 (November 2009): 328.
21. O'Brien, *Without My Cloak*, 10.
22. Ibid.
23. Ibid.
24. Rita Felski, *Doing Time: Feminist Theory and Postmodern Culture* (New York and London: New York University Press, 2000), 25–6.

25. Ibid., 3.
26. Ibid.
27. Kate O'Brien (1941), *The Land of Spices* (London: Virago, 2002), 157.
28. The O'Mara Papers: Calendar of Kate O'Brien Material. Letter from Kate O'Brien to Anne (Nance) O'Mara, March 11, 1941. 795(5). Special Collections, University of Limerick.
29. Lorna Reynolds, 'Kate O'Brien and her "Dear Native Place",' *Ireland of the Welcomes*, September–October, 1990: 35.
30. Kate O'Brien (1958), *As Music and Splendour* (London: Penguin Books, 2006), 8.
31. This reading of *As Music and Splendour* first appeared in my article 'Kate O'Brien: Writer Wanderer, Revolutionary' for the Women's Museum of Ireland. http://womensmuseumofireland.ie/articles/kate-o-brien.
32. O'Brien, 'Limerick, as I Remember It,' 51.
33. Ibid., 4.
34. Reznicek, *The European Metropolis*, 6.
35. O'Brien, *My Ireland*, 30.
36. Ibid.
37. Ibid.
38. Kate O'Brien, 'Long Distances,' *Irish Times*, April 7, 1969.
39. Laura Arrington, 'Irish Modernism,' in *Oxford Research Encyclopaedia of Literature*. Subject: British, Irish, Scottish, and Welsh Literatures. Online Publication Date: Feb 2017. https://doi.org/10.1093/acrefore/9780190201098.013.237. Accessed 17 December 2017. 12.
40. Boland, 'The Treasury of Kate O'Brien's Achievement,' 8.
41. 'Freedom of City for O'Brien,' *Cork Examiner*, May 27, 1967, 9.
42. O'Brien, *My Ireland*, 148.

References

Arrington, Laura. 'Irish Modernism.' In *Oxford Research Encyclopaedia of Literature*. Subject: British, Irish, Scottish, and Welsh Literatures, 1–19. Online. February 2017. https://doi.org/10.1093/acrefore/9780190201098.013.237. Accessed December 17, 2017.

Boland, Eavan. 'The Treasury of Kate O'Brien's Achievement.' The Women's Press. *The Irish Press*, December 12, 1970, 8.

———. 'Continuing the Encounter.' In *Ordinary People Dancing*, ed. Eibhear Walshe, 15–23. Cork: Cork University Press, 1993.

———. *Object Lessons: The Life of the Woman and the Poet in Our Time*. Manchester: Carcanet Press, 2006.

Felski, Rita. *Doing Time: Feminist Theory and Postmodern Culture*. New York; London: New York University Press, 2000.

Kemmy, Jim. 'Limerick's Lost Lady.' *New Hibernia* (May 1986): 22.

Lloyd, David. *Irish Times: Temporalities of Modernity*. Notre Dame, IN: Field Day, 2008.

Moi, Toril. '"First and Foremost a Human Being": Idealism, Theatre, and Gender in *A Doll's House*.' *Modern Drama*, 49, no. 3 (Fall 2006): 256–284.

O'Brien, Kate. *Without My Cloak*. New York: Doubleday, 1931.

———. *Mary Lavelle*. London: Heinemann. 1936.

———. 'Limerick, as I Remember It.' *Féile Pádraig Annual*, 1957, 49–51.

———. *My Ireland*. London: B.T. Batsford, 1962.

———. *Presentation Parlour*. London: Heinemann, 1963.

———. 'Long Distances.' *Irish Times*, April 7, 1969.

———. *The Ante-Room*. London: Virago, 2001.

———. *The Land of Spices*. London: Virago, 2002.

———. *As Music and Splendour*. London: Penguin Books, 2006.

O'Toole, Michael. 'Peasants to Princes.' *Management: Journal of Irish Management Institute* (December 1995): n.p.

Pearson, Nels. *Irish Cosmopolitanism: Location and Dislocation in James Joyce, Elizabeth Bowen, and Samuel Beckett*. Gainesville: University of Florida Press, 2015.

Reynolds, Lorna. 'Kate O'Brien and Her 'Dear Native Place".' *Ireland of the Welcomes*, September–October 1990, 33–35.

Reznicek, Matthew L. *The European Metropolis: Paris and Nineteenth Century Irish Women Novelists*. Clemson: Clemson University Press, 2017.

Roche, Anthony. 'The Ante-Room as Drama.' In *Ordinary People Dancing: Essays on Kate O'Brien*, ed. Eibhear Walshe, 85–199. Cork: Cork University Press, 1993.

Travis, Charles. 'The "Historical Poetics" of Kate O'Brien's Limerick: A Critical Literary Geography of Saorstát Éireann and the 1937 Bunreacht na hÉireann Plebiscite.' *Irish Geography* 42, no. 3 (November 2009): 323–341.

Walshe, Eibhear. *Kate O'Brien A Writing Life*. Dublin: Irish Academic Press, 2006.

CHAPTER 4

Migrants in the City: Dublin Through the Stranger's Eyes in Hugo Hamilton's *Hand in the Fire*

Molly Ferguson

In Hugo Hamilton's 2010 novel *Hand in the Fire*, the first Irish novel written from an Eastern European perspective,[1] Serbian immigrant Vid Ćosić is a witness to and conduit of Irish shame amidst revelations of abuses of institutional power during the Celtic Tiger period. The novel is narrated from Vid's perspective as one of the so-called new Irish—an inward migrant who arrived on Irish shores during the economic boom of 1997–2007—who experiences the conditional hospitality of Dublin.[2] During this boom, the Irish economy was rapidly expanding, with workers arriving to fill service positions and support the booming construction and renovation market that would develop into a housing bubble.[3] At first, Vid is focused on fitting in, self-conscious of his identity as a Serbian whose father was a secret police interrogator: 'what do you say when the country you grew up in can only be remembered for one thing?'.[4] He meets Kevin Concannon and his girlfriend Helen in a pub in Dublin after finding Kevin's cell phone, and shortly after Kevin puts his 'hand in the fire' for Vid by beating up an electrician who attacked Vid for kissing his daughter.[5] As Vid returns Kevin's loyalty by taking the fall with the law for his

M. Ferguson (✉)
Department of English, Ball State University, Muncie, IN, USA
e-mail: meferguson2@bsu.edu

M. Beville, D. Flynn (eds.), *Irish Urban Fictions*, Literary Urban Studies, https://doi.org/10.1007/978-3-319-98322-6_4

friend's aggression, he quickly becomes ingratiated into the Concannon family through his carpentry labour at the home of Kevin's mother, Rita. He also befriends Rita's estranged husband Johnny, a returned economic migrant. This causes Vid to become alienated from Kevin, who had disowned his father. When Vid is ultimately shunned by his new host family, they reject his presumption of insider status and are embarrassed by his knowledge of their shameful past actions, which become representative of the corruption of Dublin and Ireland more generally during the Celtic Tiger period.

Hamilton's Dublin in *Hand in the Fire* is a city of both wonder and insecurity, a space where the formerly colonised are redefining themselves by exerting power over the marginalised. As a migrant worker in Dublin, Vid's position outside of citizenship allows him to register the fundamentally unequal treatment of both migrants and women in Ireland and the reversion to colonial hierarchies that shape his exclusion. The city is figured in the novel as a site of tension between insiders scrambling for power and outsiders desiring integration and access to resources. Vid relates his own outsider status in spatial terms, as 'gate-crashing, like you're at a party and people are wondering where you came from and who invited you'.[6] As he renovates his friend's family home, seen through his eyes as a set of beams and nails, so too does Dublin emerge as a house with 'good bones', so to speak, that urgently needs renovation. Relatedly, Hamilton suggests that, despite Dublin's increasingly cosmopolitan character, gender inequities continue to structure the capital city. Both these sociopolitical and gendered forms of exclusion, I argue, produce a sense of shame pervading all relationships in Dublin specifically and the nation more generally. Hamilton's narrativisation of these distinct but related forms of exclusion and stratification are most effectively viewed through six affective public and private spaces—the home, the pub, the court, the pier, the street, and the church—that structure the city and, ultimately, the Celtic Tiger nation.

As a newcomer, Vid identifies Dublin's central contradiction as both a cosmopolitan European capital and a damaged site of colonial violence: 'The city was the greatest place on earth but it was also a dump and a war zone, depending on your entry level'.[7] The unique perspective of the migrant foreigner on Celtic Tiger Dublin reveals a rich set of cultural narratives that construct the city's emergent character through their contradictory pressures. James Holston and Arjun Appadurai write that cities are, 'the place where the business of modern society gets done'.[8] As a voyeur in Dublin, Vid becomes a witness to the flows of global capital;

the networks of gender, race, and class that align for the powerful; and the underbelly of corruption and violence that perpetuates marginalisation of the weak. As Caroline Herbert notes, the city is not only a locus for colonial oppression 'but also of anticolonial resistance'.[9] Following Herbert, I argue that Vid narratively re-makes Dublin through his observations and engagement with it and that his movements throughout the city stir up and expose the changing landscape of wealth, power, and citizenship in Ireland. Herbert also claims that 'while the postcolonial city is haunted by the spatial violence of the past, it is also marked by the exploitations and exclusions of the present: the Manichean city of colonialism reemerges in new forms in postcolonial urban space'.[10] Through Vid's narration of Celtic Tiger Dublin, the city represents a microcosm of competing desires regarding capitalism, globalisation, gender, and national identity in Ireland today.

Wandering the city as a newcomer but not a tourist, Vid initially registers the welcoming feel of Dublin as tied to its economic growth: 'My first impression here was of everything being so wealthy and inviting'.[11] He situates himself among the arguably masculine architecture of Dublin, which centralises urban commerce around sites of colonial uprising such as the General Post Office, the Four Courts, and Dublin Castle. Dublin is littered with statues that enshrine the lives of Irish revolutionaries to colonial rule, but (with the exception of Constance Markievicz) that historical role is defined as exclusively male. Women are most visible in Dublin's public spaces not on street signs or bridges, but as imagined abstractions such as Anna Livia, Queen Maeve, and Molly Malone.[12] Thus two ideologies emerge as entwined for a Serbian observer of Dublin—nationalism and masculinity. As Herbert explains, the bodies obscured by a postcolonial city's narrative of itself are 'made spectral by networks and flows of global capital'.[13] Vid's awareness of the mutual invisibility of women and immigrants like himself in Dublin's self-narrative emerges from his physical experience of navigating urban topography.

Sara Ahmed's *The Cultural Politics of Emotion* offers a framework for understanding how shame as an emotion brings the nation into being through shared coalescence of feeling. She contends that national shame recognises past abuses, while cathartically purging individual guilt over those abuses.[14] *Hand in the Fire* may be read through Ahmed's framework of national shame, because Vid brings his own ethnic disgrace as a Serbian to his observation of Irish shame, an emotion that peaked when the economic boom collapsed. Ahmed's words may be applied to Vid's role in this process: 'Shame as an emotion requires a witness: even if a subject feels shame when he or she is alone, it is the imagined view of the other that is taken on by a subject in relation to herself or himself'.[15] Vid's witnessing of gender-based Irish shame is complex, because as a white man he experiences freedoms in the city, yet as a liminal foreign migrant he identifies more with women and is receptive to their stories related to gender inequality. Therefore, Vid becomes a carrier of secrets and shame and must be excluded by the Concannons so that they may avoid shame about their family's complicity in structural violence.

Vid's words encourage a reader to imagine the migrant's cartography differently: 'And maybe I needed a different sort of map altogether', he muses, reflecting on navigating his way as a newcomer to Ireland.[16] Experiencing the novel through the metaphor of city mapping, by zooming in on six resonant sites—the home, the pub, the court, the pier, the street, and the church—allows, I argue, for a geopolitical picture of Dublin's power networks as encountered by the inward migrant. After a brief stint working security at a nursing home, Vid takes up freelance carpentry work in the private home of the Concannon family. He enters the space of the pub through his friendship with Kevin, accessing masculinity through drinking culture. Moving to the formal space of law enforcement, his encounters in the courts and questioning rooms remind him of the institutional power and violence wielded back home in Serbia. Vid's visits to the pier, a site of emigrant exits and migrant entrances, imagine it as a space that rehearses the traumas of exile. Having been expelled from the family, Vid's walking movements through the city capture it as a dangerous environment of surveillance. While always subtly present in the novel, the influence of the Church takes place at a remove from the city, in the coastal towns of Furbo and Barna. The emotional movement of Vid's narrative places emphasis on the process of being first welcomed, and then rejected, spatially depicted as a movement from the hospitality of the family home in Dublin to the alienation of the streets. David Lloyd's call for further research of Irish affective spaces in light of the nation's colonial legacy guides this inquiry. According to Lloyd, 'Much remains to be done by producing a 'gender history' of Irish social spaces and their refiguration within nationalist as well as colonial projects of modernization'.[17] Analyses of Dublin as a colonial metropolis must also be sensitive to how the city re-inscribes gendered forms of oppression.

The treatment of inward migrants in Ireland in recent years directly counters the exported tourism narrative of 'Ireland of the Welcomes' or

'*Céad míle fáilte*'.[18] Irish hospitality is a famed trait, and yet Vid's experience reveals the welcome he receives as both hollow and conditional. He initially thinks that the Concannons 'gave me this great feeling of being at home,' and he experiences a great sense of pride at being 'part of the family'.[19] However, as Jacques Derrida argues regarding hospitality, 'The Foreigner shakes up the threatening dogmatism of the paternal *logos*: the being that is, and the non-being that is not'.[20] Vid's presence draws attention to and challenges the 'paternal *logos*' of the Concannon family; a family that is significantly missing its father who emigrated to England. Derrida reminds us that 'a foreigner can be a parricide only when he is in some sense within the family,' and Vid's presence becomes threatening to Kevin as the self-appointed head of household.[21] Vid's discovery of embarrassing family secrets and his testimony of exclusion in Dublin threatens Kevin's assumed position as the patriarch by shaming him. As a foreign outsider, Vid is only accepted when he is in his assigned place as a hired worker and a sidekick for Kevin; once he demands reciprocity, he is ostracised.

From big houses to ghost estates, domestic space in Ireland has symbolically functioned in Irish literature as a feminised site of containment for corruption, violence, and shame. In theorising about the recession following the Celtic Tiger, Sinéad Molony asserts that 'intersecting discourses of gender and class have shaped representations of the Irish home as symbolic of the recessionary nation'.[22] Within this burdened ideological space of the home, women are simultaneously cast as 'the heart of the nation' and yet experience intersectional gender and class exclusion from the national collective.[23] While Vid is not a woman, as a hired migrant worker, he occupies a similarly disadvantaged position. Vid is simultaneously an employee, a welcome guest, and an intruder in both Dublin and the Concannon family home. When he begins work on the house, he describes it as a 'beautiful, spacious family home on a terraced street leading down to the sea' that was 'clearly in need of some repair', in a row of once-lovely homes Kevin now calls 'Desolation Row'.[24] Having kicked her husband out, the home is Kevin's mother Rita's empire. The first time Vid meets Rita, she looks at him 'as though I had just broken in and couldn't find my way out again'.[25] Through the 'security zones' he passes through 'with each cup of tea,' Vid learns that Rita wears a prosthetic breast, that she hid a cache of letters from her husband to the children, and that Johnny dragged the family dog to death and punched Rita before she threw him out.[26] His knowledge creates an intimacy that humiliates her, an intimacy she rescinds by warning him to finish his work: 'Don't hover'—a message many inward migrants received as the Celtic Tiger economy dwindled.[27]

Vid does a series of carpentry jobs for the Concannons from building an armoire to replacing a door and lock, the last of which is completely unnecessary—turning the floorboards around to run lengthways instead of vertical. He attempts to refuse Rita's request on the basis that it is a wasteful, expensive endeavour: 'All the floorboards taken up. All the joists underneath would have to be turned around. There would be a lot of wastage, new joists, broken floorboards to be replaced. All the noise. Sawdust all over the house'.[28] Rita is not to be denied this unnecessary and extravagant alteration to the very floor beneath her feet, enacting a predictable stereotype of gendered consumerism, a perception that during the Celtic Tiger Irish women spent money extravagantly for luxuries, leading to the recession.[29] Vid's employment is a way for the Concannons to ensure that he will put his 'hand in the fire' for Kevin by taking the blame in court for Kevin's assault of the electrician.[30] Kevin gives Vid a key and declares him family, and Vid imagines himself 'receiving a certificate, rolled up with a ribbon, stating that I belonged there'.[31] As an immigrant, Vid's 'certificate' of belonging legitimises him, yet since his value is conferred only through a paid labour relationship, it remains superficial. Just as the home renovations mask the secrets within, so too do they obscure the exclusion of inward migrants trying to make their homes in Ireland's capital.

The Concannon home acts as a symbol for Dublin during the Celtic Tiger period. Like Celtic Tiger Dublin, the house has 'great charm' but is 'a big, draughty place' that needs energy efficiency repairs rather than the more glamorous floorboard-turning.[32] Outward show, it seems, is being valued in this home rather than long-term improvement of the structure against the wind and rain. In extended symbolism of Celtic Tiger Dublin, the house has 'the touches of mould, black patches left behind the book case, at the backs of the sofas'.[33] Though not immediately apparent to the observer, a spreading mould of corruption will end the economic prosperity in Ireland, just as the Concannon house is neglecting its own creeping family disease in favour of cosmetic renovation. Kevin's shameful compulsion to contain his drug-using, pregnant sister Ellis from 'ripping loose from the family' further reveals the rotting mould within the externally lovely home, permeated with a horror of women's bodies.[34]

Acting as an alternative city space to the home and work, the pub is a space of male bonding that excludes women and minorities. According to

Lloyd, the pub 'constitutes a third term' that acts as a 'rival to the home in providing an alternative space for male conviviality, leisure and community, one not yet subordinated to the regulations of private domesticity and, accordingly, 'public'.[35] The pub is a site of male dominance where women's sexuality belongs to men. For example, the electrician views his daughter as his property and becomes enraged when Vid, a foreigner, kisses her. In another pub scene Vid releases his Moldovian girlfriend Liuda (who has been sexually exploited by an employer who let her work permit lapse) to an Irish-born man who excuses himself to Vid that he's 'just having the craic'.[36] Vid's participation in Irish drinking culture illuminates its dangerously predatory environment through an outsider's eyes, and he conflates his inability to drink and have casual sex with his assimilation as a new Irish resident: 'I was unfit to integrate. I couldn't get drunk. I couldn't laugh'.[37] Since he perceives integration with native Irish to require excess drinking and sexual aggression, he remains outside of Irish masculinity as conceptualised in the novel.

As a foreigner, Vid is considered not fully masculine by Kevin, yet their greatest moments of closeness are bonding through 'lad culture', via the centrality of pubs in Dublin's social culture. Their primary mode of intimacy is drinking together, and Kevin uses Vid as a cover for cheating on his girlfriend, Helen, even forcing him to watch. After one drinking session, Kevin advances sexually on Vid, thrusting his tongue in Vid's mouth and then immediately disavowing the act. Vid pretends to use Kevin's drunkenness as an excuse for his lapse in heterosexuality, but Kevin nevertheless holds Vid's knowledge of his 'secret' against their friendship. Kevin combines sexually charged language with ethnic slurs towards Vid when he realises that Vid has been meeting Kevin's estranged father Johnny, calling him 'You fucking Serbian cunt', 'little Balkan creep', and 'Euro-fuck'.[38] Vid's understanding of the family's humiliation at casting off their father is read by Kevin as an attempt to mix friendship (the pub space) with family (the home), and he attacks Vid's ethnicity and gender to remind him that he is alien to how these spaces operate. The pub is also where Vid realises his friendship with Kevin has ended, signified by three untouched pints of Guinness left on the bar when Kevin and his new friends leave after seeing Vid there. Given that pub culture, notably in Temple Bar, is a highly visible marker of Dublin's commercial brand as a global city, Kevin's abandonment of Vid in the pub denies him the *craic* that signifies belonging.

While he can blend in to the pub space, Vid is most uncomfortably alien in the institution of the court and in police questioning rooms, which he confronts when he must defend himself against the charge of assaulting the electrician, though Kevin committed the crime. Once again, the external framing of the institutional space maps on to the affective relationships in the text. The friendship between Vid and Kevin becomes coded in legal and economic language in *Hand in the Fire*, suggesting that their friendship was a transaction rather than an emotional connection. As Vid explains, 'The rules of friendship included a code of client confidentiality'.[39] The slippage from friend to client is not accidental, as Vid is indeed Kevin's client at the trial. In court, Vid reacts to the symbol of the harp on the wall by internalising Irish shame that 'the Irish were now doing things to themselves that no oppressor would ever have dreamed of'.[40] Such a loss of national ideals is made vividly clear through the international Dublin cityscape, which is burying its memory of colonial dominance by becoming increasingly hierarchical and economically exploitative, while simultaneously maintaining outmoded gender inequities. Further, Hamilton demonstrates how these regressive tendencies produce a pervasive sense of shame felt throughout the city and demonstrated through Vid's experiences in a range of affective public and private urban spaces.

For example, Vid frequently wanders to the pier on his many walks throughout Dublin, a liminal space that is the point of exit via emigration and entrance via immigration, and is figured in *Hand in the Fire* as a site of trauma. Trauma is an essential ingredient of the encounter between the inward migrant and the Irish host, a wound that reflects both the migrant's past and the Irish collective memory of emigration and exile. Ronit Lentin's position is useful here; she contends that the traumatic memory of emigration is repressed in Irish experience but comes to the surface in encounters between the inward migrant and the Irish host.[41] Lentin explains how the traumatic past of emigration returns to haunt the Irish, or as she calls it the 'return of the repressed,' a collective memory of emigration the Irish must address in order to interact positively with inward migrants.[42]

In Vid's walks through Dublin, he experiences the harbour and the Carlisle pier as a porous border for emigration and immigration, framing the interactions he witnesses through postcolonial and global pressures to move from place to place. He goes to the pier while looking for Ellis who has run off on a drug bender, and it reminds him of a 'grey, derelict arm sticking out' and a 'derelict monument to emigration'.[43] During the Celtic

Tiger period, the pier is a juxtaposition of new wealth and those it left behind, with the ticket booth now a yacht sales agency and the dock now a super ferry terminal, while the space itself is occupied by the homeless, gangs, and drug addicts. These transients, like the 'knackers' (Travellers) and 'spongers' (people on social welfare) he hears the builders rail about in the pub, are the bottom rung of Dublin society whom Vid recognises as being vulnerable and unwanted like himself. Vid thinks, 'The druggies are the real exiles now,' who function 'like a separate ethnic group' at the edges of the city, including users like Ellis, whom he terms 'had emigrated to the land of dreams and drugs'.[44] The pier also functions as a recuperative space of memory for exiles. Johnny remembers purchasing a '99' cone at Teddy's ice cream shop in Dún Laoghaire, a bittersweet memory that triggers shame as an absent father, prompting him to send his children money annually for a vicarious experience of this quintessential Dublin pleasure. Vid's empathetic witnessing of pain experienced at the pier offers the porous border of emigration and immigration as a spatial depiction of disadvantaged groups being pushed to Dublin's outer edge in the contemporary neoliberal capitalist ethos.

In Dublin's public spaces, Vid enjoys mobility as a white man, and yet his narrative is informed by paranoia reflecting the novel's gradual movement from hospitality to surveillance—a movement reflective of Dublin's transformation from an inclusive Irish space to a globalised city anxious about its changing character. In his interview with Aisling McKeown, Hamilton identifies a perception that immigrants are 'overhearing us on the streets', resulting in the Irish needing to reflect on their speech and comport themselves differently. The author adds, 'I think there's something interesting happening in Ireland at the moment, we're being examined'.[45] While Vid is observing the Concannons and making assessments about them and about Irish people in general, he himself is under constant scrutiny as an immigrant. Herbert's description of the postcolonial city as a 'key site for policing the colonizer and the colonized' is made manifest in Vid's movements in the working-class parts of Dublin.[46] At the building supply store where he meets the other labourers, he recognises it as a site of observation, noting that 'it was not entirely about getting timber and plumbing needs but also a place where people kept an eye on each other'.[47] Through Vid's experiences, Hamilton thus captures a Dublin in transition, one increasingly interested in economic advantage but also one teeming with paranoia and shame about trading its colonial resistance for capitalism.

The movements of those who do not belong are recorded by the mechanical surveillance of security cameras in the urban environment. Returning to Herbert, in the postcolonial city, the 'deployment of the technologies of power' allows the coloniser to regulate its subordinates.[48] Despite decolonisation and modernisation, these characteristics remain salient in contemporary Dublin as Vid witnesses it. Kevin installs cameras at his mother's house after an arson attempt, and the cameras at the boatyard both result in Johnny's death and become Vid's alibi when the thugs sent by the electrician chase him there. Throughout the novel, Vid is often looking to 'make sure that I was not being followed', and he experiences paranoia at the thought that 'people had begun to check me out on the internet and rake up my history' to blame him for the human rights abuses in Serbia.[49] As a marked foreign body, Vid's awareness of being watched reveals the technologies of surveillance in the modern postcolonial city. He manipulates mechanical surveillance equipment to his own advantage when the electrician's friends pursue him, running to the yacht club where Johnny works and using the security cameras to signal to Johnny for help, which was 'recorded for later use in evidence' that exonerates him when he refuses to speak to police.[50] Contemporary cities are sites of panoptic surveillance, and increased observation in Dublin marks a shift towards emulating London, where recent estimates state that approximately 500,000 CCTV cameras collect daily footage of people in public spaces.[51] The slippage from hospitality to surveillance in the postcolonial city is illustrated in the novel as a spatial movement from the home to the streets, a rejection that characterises the urban immigrant as a precarious body.[52]

Like the associative thought process of Joyce's Leopold Bloom in *Ulysses*, Vid's walking through Dublin reveals much about its gendered and nationalist architecture. Dublin's literary landmarks, though sites of cultural heritage and national pride, are alienating to Vid as a stranger to the city. While tourists may enjoy visiting these monuments to Irish dominance in literature, Vid's feelings of disconnection from cultural landmarks in Dublin express the sense of exclusion these markers convey to non-native Irish people. For example, he takes Liuda on a date to Howth, finding that neither of them is interested in Joyce's work or the significance of the place, with no context for comprehending it other than that Joyce first had intercourse with his future wife Nora there and that the Irish continue to celebrate it on Bloomsday. He describes Howth cynically as a bland commercial landscape devoid of memory: 'just another hill, basically, with a big golf course and some wealthy villas and gates and planes landing nearby at the airport'.[53] In realising that it was a mistake to bring Liuda there and that their relationship is fizzling out, he admits, 'it already belonged to somebody else. We were the latecomers'.[54] Vid's symbolic impotence with his lover via Dublin's literary history casts him once more as simultaneously foreign and emasculated.

In urban streets, vehicles are sites of potential violence, and Vid is most frequently walking, exempt from the brutality that happens within them. Vid witnesses Kevin's aggressive masculinity when he urinates on Helen's windshield in front of Helen and himself, causing him to question: 'Was he showcasing his power over her? Over me?' (32). Helen's powerlessness as a woman is leveled with Vid's stunned viewing as an immigrant here. The scene with Helen and Kevin is mirrored later when Vid happens to see a man and a woman pulled over arguing in a car, which escalates into the man beating the woman so badly that when he drives off Vid is unsure she survived. The disturbing repetition of this scene with Kevin's earlier display of dominance is evident when the man urinates next to Vid in an alley before getting back in the car. Thus, Vid's perspective reveals Kevin through Molony's profile of the Irish man as a trope in Celtic Tiger discourse: 'this figure of virile entrepreneurial masculinity functions as an ideal neoliberal body around which all other modes of living must adjust'.[55]

It is telling that in Hamilton's affective spatial matrix, the church is displaced far from the city out to the West coast and islands. In fact, the story of Kevin's paternal aunt Máire Concannon, a woman from Furbo in Connemara who either drowned herself or was drowned by her parish, acts as a code for unlocking Irish trauma and Vid's own need to remember his past. The woman becomes pregnant out of wedlock and is violently excluded from her community through the religious authority of the priest. As Kevin relates the story to Vid, the priest declared: 'if the men in the area were not men enough to drown her, then perhaps she would have the decency to drown herself'.[56] Her pregnant body washes up on the shore of the Aran Island of Inishmore shortly after. Vid is aware that the horror of silence in this story has created a 'darkness' in the family, but beyond family legend it has become folklore on the island, where 'Something unspeakable about her death [entered] into the memory of the landscape'.[57]

Vid leaves Dublin, retracing the woman's steps from Furbo to Barna in a pilgrimage to get answers from churches in both towns about her death on behalf of the Concannons, but also to confront the tribal violence in his own past. He identifies with the woman, a mirroring Hamilton suggests is equivalent with his migrant status: 'her residence permit had run out, you might say [...] she had lost her rights and become an alien'.[58] He thinks, 'The story Kevin told me seemed not unlike the way the secret police operated in my own country and the paralysis that people felt in the face of authority'.[59] When complicity with the Church's violent shaming of women is posited as equivalent with Serbians amidst genocide, Hamilton suggests that human rights abuses against women in Ireland signal a deeper institutional violence—one underwriting actions in the metropolis but whose seats of power are strategically sequestered at the edges of the cityscape, removed from surveillance.

The national shame for how women were treated by the Catholic Church is affectively repeated in contemporary treatment of inward migrants in Ireland by the state, which Hamilton makes evident through Vid's experiences in Dublin. Vid not only listens to Máire Concannon's story told by three different characters but also hears the story of his co-worker Bridie who was forced by 'savages' (the nuns) to give up her child, and the story of a painter who was given up as an illegitimate child and remained a family secret as an adult, meeting 'undercover' with his mother after discovering her.[60] The church, though relegated to the furthest point from Dublin or 'driven off the map and into the sea,' nevertheless exerts a violent and oppressive force that continues to shame the Concannons and the Irish at large.[61] Vid carries the shame of the woman's descendants for complicity in her death back to Dublin with him, exposing it to light for assessment and healing.

Finally, Vid's relationship with the Concannon home, and by extension, the city, comes full circle when it is his 'very good information about the house' that allows him to re-enter even after he has turned in his key, eluding the new security cameras and lights and tiptoeing around the remembered creaks in the floorboards, to 'rescue' Ellis from Kevin and Rita who have booked her passage to England for an abortion.[62] History seems about to repeat itself with Ellis pregnant by her drug-addicted boyfriend and Kevin threatening to disown her, but after Vid and Helen intervene, he loosens his grip, reassuring her, 'Whatever you decide is right'.[63] To Vid, this act of freeing Ellis to make her own decision about her pregnancy is 'where I entered into the story of the country at last,' declaring his agency instead of 'letting things happen around me as if I was still only an immigrant and it was none of my business'.[64] Though the house, like Dublin, initially promised a welcome, trespassing ironically becomes the means to claim his place in Ireland. That Vid, a man but marginalised, symbolically earns his citizenship by ensuring Ellis' reproductive rights suggests that gender inequality fundamentally structures the axis of national identity. As Dublin becomes increasingly culturally porous via inward migrants, Hamilton suggests that their presence may question the underpinnings of a nationalism dependent upon gender-based oppression.

Anne Enright's review of *Hand in the Fire* characterises Hamilton's writing as 'the voice of the migrant, the mongrel, of the person who is neither one thing nor the other, of the stranger and the traveller in us all'.[65] In the case of Ahmed, shame may be brought onto the nation through 'illegitimate others' who incite shame because they 'cannot reproduce the national ideal'.[66] While shame allows the nation to witness past injustice 'as it exposes the failure of the nation to live up to its ideals,' it also allows for movement forward by reckoning with injustice and working towards a nation that fits the collective imaginary.[67] Edna O'Brien's famous pronunciation of Ireland as 'a land of shame, a land of murder, a land of strange, throttled, sacrificial women' holds true in this novel,[68] where the traumatic secrets of Ireland's past are revealed as embedded in structural violence against women. When Vid's trajectory through Dublin as an inward migrant is mapped onto a movement from the house to the streets and beyond, his ability as an outsider to reveal the gendered exclusions of Dublin as a postcolonial city becomes salient.

Notes

1. Enright, '*Hand in the Fire*'. n.p.
2. *Hand in the Fire* appears to be set after large numbers of inward migrants began arriving to fill labour gaps in Ireland, but likely before the peak of backlash against immigrants that led to the '*jus soli*' referendum in 2004 (see Fanning and Mutwarasibo). Also in 2004, ten countries, seven of which were either former Soviet republics or satellites, joined the European Union (EU) and were allowed to move freely as EU citizens and access social welfare in member states. Ireland's 'open-door policy' was not shared by the rest of Western Europe aside from Sweden, but the nation's expanding economy welcomed the newcomers, in part, because they were European and therefore fit in to its self-image of a progressively multicultural, albeit white, state (see Villar-Argáiz and Lentin). Another effect of EU enlargement was that Ireland moved away from its more open work permit system, choosing instead to have the new EU population fill its need for low-skilled and manual labour. From 2004 to 2007, migrants

from EU member states came in increasing numbers, often taking jobs below their education or skill level (see Ruhs and Quinn).

3. While Serbia is not yet in the EU, Vid most likely obtained what he calls his 'residence permit' under the old system that was more inclusive of non-EU migrant workers (84). According to Ruhs and Quinn, between 1999 and 2003, there was a 650% increase in non-EEA work permits issued, a number that sharply dropped off after EU enlargement in 2003.
4. Hamilton, *Hand in the Fire*, 2. Vid is mistaken for Polish by the electrician, probably because the Polish were the largest group of Eastern European migrants to enter Ireland through the EU expansion (see Ruhs and Quinn), and he chooses to prolong that false impression of his ethnicity to avoid being associated with war crimes committed by Serbians like his father.
5. Ibid., p. 30.
6. Hamilton, *Hand in the Fire*, p. 19.
7. Hamilton, *Hand in the Fire*, p. 216.
8. Holston and Appadurai, 'Introduction', p. 3.
9. Herbert, 'Postcolonial Cities', p. 201.
10. Ibid., p. 207.
11. Hamilton, *Hand in the Fire*, p. 53.
12. The Rosie Hackett bridge, the only bridge in city centre named for a woman, opened later, in 2014.
13. Herbert, 'Postcolonial Cities', p. 203.
14. Ahmed, *Cultural Politics*, p. 101.
15. Ibid., p. 105.
16. Hamilton, *Hand in the Fire*, p. 6.
17. Lloyd, *Irish Culture and Colonial Modernity*, 91. Lloyd is referring here to early twentieth-century Dublin, but his identification of 'gendered social spaces' also suits my reading of Hamilton's novel.
18. The Irish tourism industry has traditionally branded Ireland as a welcoming place. The phrase *Céad míle fáilte*, which means 'one hundred thousand welcomes', is displayed on souvenirs from dish towels to salt-and-pepper shakers.
19. Hamilton, *Hand in the Fire*, p. 30.
20. Derrida, *Of Hospitality*, p. 5.
21. Ibid., p. 7.
22. Molony, 'House and Home: Structuring Absences', p. 184.
23. Ibid., p. 187.
24. Hamilton, *Hand in the Fire*, p. 16.
25. Ibid., p. 17.
26. Ibid., p. 162.
27. Ibid., p. 218.

28. Ibid., p. 68.
29. Emerson, Newton. 'Working Women Almost Certainly Caused the Credit Crunch'. *Irish Times.* February 2, 2009.http://www.irishtimes.com/opinion/working-women-almost-certainly-caused-the-credit-crunch-1.709198. Emerson's satirical piece claimed that 'women were the driving force behind the greed, consumerism and materialism of the Celtic Tiger years and it was female employment that funded their oestrogen-crazed acquisitiveness'. Many readers did not read the piece as satire and agreed wholeheartedly with its premise in online comments, causing controversy.
30. Hamilton, *Hand in the Fire*, p. 41.
31. Ibid., p. 69.
32. Ibid., p. 74.
33. Ibid.
34. Ibid., p. 141.
35. Lloyd, *Irish Culture and Colonial Modernity*, p. 105.
36. Hamilton, *Hand in the Fire*, p. 63.
37. Ibid., p. 105.
38. Ibid., p. 126.
39. Ibid., p. 78.
40. Ibid., p. 94.
41. Lentin, 'Anti-racist Responses', p. 233.
42. Ibid.
43. Hamilton, *Hand in the Fire*, p. 183.
44. Ibid., p. 181, 181, 185.
45. McKeown, '*Hand in the Fire* and the Bid to Belong'.
46. Herbert, 'Postcolonial Cities', p. 200.
47. Hamilton, *Hand in the Fire*, p. 154.
48. Herbert, 'Postcolonial Cities', p. 201.
49. Ibid., p. 220, 224.
50. Ibid., p. 247.
51. McCahill and Norris. 'Working Paper no. 6: CCTV in London'.
52. See Butler, Judith. *Precarious Life.*
53. Hamilton, *Hand in the Fire*, p. 59.
54. Ibid., p. 59.
55. Molony, 'House and Home: Structuring Absences', p. 185.
56. Ibid., p. 88.
57. Ibid., p. 89. The woman's place of death also retains her story in its name: *Bean Bháite*, or 'drowned woman' (262).
58. Ibid., p. 173.
59. Ibid., p. 89.
60. Ibid., p. 13, 218.

61. Ibid., p. 210.
62. Ibid., p. 262.
63. Ibid., p. 270.
64. Ibid., p. 261.
65. Enright, '*Hand in the Fire*'.
66. Ahmed, *The Cultural Politics of Emotion*, p. 108.
67. Ibid., p. 109.
68. O'Brien, Edna. 'A Scandalous Woman', p. 265.

References

Ahmed, Sarah. *The Cultural Politics of Emotion*. New York: Routledge, 2004.

Butler, Judith. *Precarious Life: The Powers of Mourning and Violence*. New York: Verso, 2006.

Derrida, Jacques, and Anne Dufourmantelle. *Of Hospitality (Cultural Memory in the Present)*. Translated by Rachel Bowlby. Stanford, CA: Stanford University Press, 2000.

Emerson, Newton. 'Working Women Almost Certainly Caused the Credit Crunch.' *Irish Times*. February 2, 2009. http://www.irishtimes.com/opinion/working-women-almost-certainly-caused-the-credit-crunch-1.709198.

Enright, Anne. '*Hand in the Fire* by Hugo Hamilton.' *The Guardian*, April 16, 2010. https://www.theguardian.com/books/2010/apr/17/hand-fire-hugo-hamilton-review.

Fanning, Brian, and Fidele Mutwarasibo. 'Nationals/Non-Nationals: Immigration, Citizenship and Politics in the Republic of Ireland.' *Ethnic and Racial Studies*, 19, no. 3 (March 2007): 439–460.

Hamilton, Hugo. *Hand in the Fire*. London: Fourth Estate, 2010.

Herbert, Caroline. 'Postcolonial Cities.' In *The Cambridge Companion to the City in Literature*, 200–215, ed. Kevin McNamara. Cambridge: Cambridge University Press, 2014.

Holston, James, and Arjun Appadurai, 'Introduction.' In *Cities and Citizenship*, ed. James Holston, 1–20. Durham, NC: Duke University Press, 1999.

Lentin, Ronit. 'Anti-racist Responses to the Racialization of Irishness: Disavowed Multiculturalism and Its Discontents.' In *Racism and Anti-racism in Ireland*, ed. Ronit Lentin and Robbie McVeigh, 226–239. Belfast: Beyond the Pale, 2002.

Lloyd, David. *Irish Culture and Colonial Modernity 1800–2000: The Transformation of Oral Space*. Cambridge: Cambridge University Press, 2011.

McCahill, Michael, and Clive Norris. 'Working Paper no. 6: CCTV in London.' *Urban Eye*. 5th Framework Programme of the European Commission. June 2002. http://www.urbaneye.net/results/ue_wp6.pdf.

McKeown, Aisling. '*Hand in the Fire* and the Bid to Belong: An Interview with Hugo Hamilton.' *Irish Studies Review* 19, no. 4 (November 2011): 427–432.

Molony, Sinéad. 'House and Home: Structuring Absences in Post-Celtic Tiger Documentary.' In *Gendering the Recession: Media and Culture in an Age of Austerity*, ed. Diane Negra and Yvonne Tasker, 181–202. Durham and London: Duke University Press, 2014.

O'Brien, Edna. 'A Scandalous Woman.' In *A Fanatic Heart*, 239–265. New York: Farrar, Strauss, and Giroux, 1984.

Ruhs, Martin, and Emma Quinn. 'Ireland: From Rapid Immigration to Recession.' *The Online Journal of the Migration Policy Institute*. September 1, 2009. http://www.migrationpolicy.org/article/ireland-rapid-immigration-recession.

Villar-Argáiz, Pilar. 'Introduction: The Immigrant in Contemporary Irish Literature.' In *Literary Visions of Multicultural Ireland: The Immigrant in Contemporary Irish Literature*, ed. Pilar Villar-Argáiz, 1–36. Manchester: Manchester University Press, 2014.

Zamorano Llena, Carmen. '"Our Identity Is Our Own Instability": Intercultural Exchanges and the Redefinition of Identity in Hugo Hamilton's *Disguise* and *Hand in the Fire*.' In *Literary Visions of Multicultural Ireland: The Immigrant in Contemporary Irish Literature*, ed. Pilar Villar-Argáiz, 108–119. Manchester: Manchester University Press, 2014.

CHAPTER 5

Phantasmal Belfast, Ancient Languages, Modern Aura in Ciaran Carson's *The Star Factory*

Tim Keane

Mapping the Unreal

Studying a map of Belfast published in 1811, Ciaran Carson notes that the city's cartographer, John Dubourdieu, has diagrammed streets that never existed and a bridge over the River Lagan that was never built. Recounting an attempt to see the River Farset as a whole entity from behind the imprisoning 'bars' of a grammar school, Carson sees the waterway in collaged fragments. 'I did not know [the river's] name, then, but was mesmerised by its rubbish: a bottomless bucket, the undercarriage of a pram, and the rusted springs sticking out of the wreck of a sunk abandoned sofa.'[1] In another context, an adult's appraisal of an inherited chesterfield sofa yields to its allegorical childhood function as a 'flying machine,' modelled on vehicles from folktales and science fiction. Belfast seems a sound stage for an unfolding *picaresque* film. Writing about a stretch of half-developed land, Carson observes 'various stages of field, leftover landscape, vacant ground, plot, building site, half-built houses, and completed semi-detacheds.'[2] Political conflict intrudes, of course, introducing additional unrealities. An

T. Keane (✉)
BMCC, City University of New York, New York, NY, USA
e-mail: tkeane@bmcc.cuny.edu

M. Beville, D. Flynn (eds.), *Irish Urban Fictions*, Literary Urban Studies, https://doi.org/10.1007/978-3-319-98322-6_5

older version of Carson visits Smithfield Market after its incineration following a guerrilla bombing in the mid-1970s to find that the shopping arcade he had enjoyed visiting in after school hours resembles a site in a post-Second World War city, 'a minor Dresden, with only the stone piers of its gates still standing.'[3] Past, present, and future blur. Carson surveys an old Parliamentary report about Belfast's riverside land use that 'sings of leisure purposes, velodromes and pleasure parks,' most unrealised, surviving as vividly documented fantasies by long-dead city planners, their maps presaging the later industrialisation of that same site, pocked with 'ruinous Gasworks' and unspooling 'cul-de-sacs and ring roads' wherein 'the city consumes itself,' all leading to Carson's aporia that '*The city is a map of the city.*'

As a swarming commercial and socio-political dynamism, filled with innumerable lifestyles and subcultures, multitudinous infrastructures and peripatetic crowds, Belfast, like any modern city, remains impervious to a definitive or singular literary representation. Its heterogeneity is the source of its apparent unreality, 'the ongoing, fractious epic that is Belfast.'[4] The phenomenon is not unique to Belfast and the challenge is not only Carson's. In *Being Modern Together*, Irish critic Denis Donoghue posits that, starting in the mid-nineteenth century, as the modern metropolis exerted its (still-unprecedented) overpowering cognitive and sensory overload on human consciousness, certain writers, starting with Edgar Allan Poe and Charles Baudelaire, sought to explore that ineffability and its baffling parts. However, the disconnected elements and infinitesimal dimensions of large-scale urbanism necessitated corresponding transformations in literary form, textual innovations conducive to the city's socio-psychological challenges, a predicament that Donoghue formulates as 'the friction between the individual mind and appearances—city streets, crowds, anonymity—over which it [the individual mind] had no control.'[5] From the blind alleyways and dimly lit flats of Arthur Conan Doyle's crime-ridden, Victorian-era London to George Perec's maze-like apartment buildings in post-war Paris, much vanguard literature within any given era represents the city as discontinuous, opaque, and inchoate, resembling, in lesser and greater degrees, what T.S. Eliot calls an 'unreal' site.[6] Indeed Eliot's infamous adjective about the London's population shows a poet blanching in the face of the ghostly quality within the city's blunt-force day-to-day commuting. Such fictive or dreamlike components define urban experience, an intimidating reality that, as Italo Calvino puts it, '[c]ities, like dreams, are made of desires and fears [...] the thread of their discourse is secret, their rules are absurd and everything conceals something else.'[7]

Commenting on Carson's corresponding determination to handle his native Belfast's cryptic mutability, poet Paul Muldoon, Carson's slightly younger peer in the Northern Irish literary scene, says he 'take[s] in the rubble and rumpus and riddling of day-to-day life in Belfast [...] taking over Belfast even as he takes it on.'[8] To 'take on' and 'take over' Belfast in *The Star Factory*, Carson relies on two seemingly divergent aesthetic and philosophical stratagems, one based in pre-urban oral Irish storytelling traditions and the other culled from modern Continental theories about cities and writing. Though rooted in different epochs and diametrically opposed worldviews, these epistemic poles are interrelated. Both forego realist or naturalistic conceptions of time and space, accommodating Carson's determination to revive antique, archaic, or otherwise neglected forms of knowledge to make sense of the modern metropolis. Writing about Belfast involves recapturing and documenting gradations and subtleties in past and present acts of perceiving the city, realities ignored through the habitual comforts of conventional realism and the distortions of chronologically oriented memory:

> Conscious perception is only a fraction of what we know through our senses. By far the greater part we get through subliminal perception [...] What do we remember of ourselves? A few fleeting fragments, which we make into the shifting histories of ourselves [...] [and] through language we make up a fictive self, we project it back into the past, and forward into the future, and even beyond the grave. But the self we imagine surviving death is a phantom even in life. A ghost in the brain.[9]

Applied to how the self identifies with its surroundings, this phenomenological project validates Carson's uncertainty principle stipulating that, within Belfast, 'everything is contingent and provisional' including its 'plan of might have beens.'[10] He compares this instability about mapping the city to the 'frottage effect' in writing, in which '[W]hole segments of the/ map have fallen off.'[11] In response to this impasse, Carson's literary representations of Belfast foreground contested, irresolvable debates between truth and error, memory and imagination, fact and fiction, advancing a radical scepticism about texts that purport to represent a city.

> No don't trust maps, for they avoid the moment: ramps, barricades, diversions, Peace Lines. Though if there is an ideal map, which shows the city as it is, it may exist in the eye of that helicopter ratcheting overhead, its search-

> light fingering and scanning the micro-chip deviations: the surge of funerals and parades, swelling and accelerating, time-lapsed, sucked back into nothingness by the rewind button.[12]

The flagship text in Carson's lifework about Belfast is *The Star Factory* (1997). The memoir transposes Irish oral traditions in storytelling into densely compacted, contemplative reportages about growing bewildered in a technology-driven, working-and-middle-class neighbourhood in a sectarian city. Carson augments these imaginative and mythic narrative Irish modes for relating urban experiences with abstemious fact-finding, civic sourcebooks and copious fieldwork. Complicating matters, the text filters childhood memories through the epistemic frameworks of phantasmagoria and aura, interrelated concepts derived from early twentieth-century German and French theories about urban culture.[13] *The Star Factory* interpolates extended paraphrases and quotations from genealogical and etymological explications about places, found objects, and the city's competing taxonomies about itself. The narrative detours and formal and structural gamesmanship parallel the zig-zagging energy, fleeting vistas, and elliptical motion within the daily bustle of Belfast. As critic Neal Alexander puts it, Carson's Belfast is, 'characterised by perpetual change, a ceaseless interplay of disintegration and construction through which [...] place is conceived not in terms of certainty and stability but as a process of dislocation and appropriation through which meanings are assembled and contested.'[14] Carson compares his city's resistance to representation to the 'frottage effect' in writing, in which '[W]hole segments of the/map have fallen off.'[15]

Archaic Modernism: Irishness, Orality, and Belfast

In 2009, following the publication of Carson's *Collected Poems*, a profile about the poet in a prominent British newspaper noted that his abiding interest in writing about Belfast inevitably invites comparison to James Joyce's literary preoccupations with Dublin.[16] Setting aside the implication of intramural Irish competitiveness and a dubious "anxiety of influence" of Joyce upon Carson, these two writers approach to Irishness and the English language through their urban writings and thus illuminate common concerns and crucial distinctions about language, storytelling, and national identity.

Through remarkably similar obsessiveness and excruciating attention to physical details, Joyce and Carson capture their respective native cities during fixed historical periods scarred by civil strife and crises about Irish nationhood. In writing about the Irish city, each draws upon their dexterity with languages other than English while also changing the rules and norms that govern the traditionally British literary genres that both writers subvert.

In reconnoitring Dublin, Joyce's novels reconstruct the English language, including diction, grammar, and syntax, partly as a literary manifestation of his psychological revolt against repressive institutional forces in modern Ireland. Examining these perturbed undercurrents, Seamus Deane theorises that Joyce's novels, immersed in Dublin, a city marked by 'paralysis,' showcase a writer trapped between unsuitable choices for claiming an Irish identity. 'Joyce rewrote the idea of a national character,' Deane notes, 'and replaced it by the idea of a character in search of a nation to which he (Joyce-Stephen) could belong.'[17] According to Deane, Joyce's quest calls for a resistance to framing personhood within a new, banal, and compromised Ireland—an Anglicised Dublin of the 1910s and 1920s—while simultaneously rejecting identification with a cliché-ridden Irish past—'Mother Ireland' and various Celtic Twilight motifs. To resolve this impasse, Joyce fabricates an alternative mode, an imaginative morphology that Deane describes as the novels' invented, phantasmal 'language of emancipation' epitomised in a mythopoetic orality, especially the ecstatic Molly Bloom and the effusive polyglot torrents of Anna Liva Plurabelle. Deane argues that Joyce's synthetic, poeticised, and retooled English, however fabricated or 'unreal' on its surfaces, is, in fact, undeniably *real* because that recalibrated language responds to inescapably real and historically bound circumstances unique to Joyce's existence in Dublin. His fiction weighs contemporaneous options for claiming Irishness and dramatises their insufficiencies, and the urban narrative rebels against these alternatives, forging a libidinous 'new' English language that upends literary convention and realises the phantasmal of Dublin city:

> [T]he escape from national character into a version of identity, or the escape into a newly constituted version of national character, is consistently mediated through the recourse to the phantasmal. The phantasmal subject and the phantasmal territory do not become substitutes for something actual. Their virtuality is the consequence of the analysis of the actual; the real subject and the real country are, in Irish conditions, representable only as the unreal—the unreally real.[18]

In writing of Belfast, Carson shares Joyce's deep distrust about the presumed authority of the English language as a vehicle for naming, controlling, and knowing an Irish city. In contrast to Joyce's approach, Carson's writing about Belfast makes no attempt to re-create the English language from within. While Carson's texts raid the hidden corners of the English lexicon, especially for its specialised terminologies, technical jargon, and archaic and polysyllabic diction, Carson does not alter English as such. Instead Carson's texts confront the fictive-and-real or, to borrow Deane's phrase, the 'unreally real' dimensions of Belfast's Anglicised Irishness through an avowed philosophical unease about language and interpretation *on any level*, an interpretive wariness so extreme and so ubiquitous that this agnosticism about meaning is a *modus operandi* throughout his *oeuvre*.

In biographical contrast to Joyce, Carson's first language was Irish, not English. Joyce's alienation from English, a metaphorical conceit cultivated by Joyce, has been advanced for decades by biographers and critics; in contrast, Carson's relationship to English is marked by an irrevocable yet constructive alienation. Born in 1948 to Catholic parents who met while taking Irish-language classes, Carson was raised as an Irish speaker in various neighbourhoods in West Belfast, learning English first on the streets and only later in school. Throughout the years, in editorial interviews and remarks to audiences, he has reaffirmed an uncertainty about all language and meaning, an abiding doubt that stems from this Irish-Anglo linguistic divergence. 'I was brought up bilingually,' he says, 'always uncertain of what I was saying because to say one thing in Gaelic, means a totally different thing in English. To say that the words are true, to the extent that they are actual expressions of a real event, to me is very uncertain.'[19] These foundational remarks about being 'uncertain' about the 'real event' shape his literary approach to Belfast. 'You can never explain the world,' he adds, 'You can only go halfway.'[20] In that same context, what Carson claims for his poetry collection *The Twelfth of Never* (1999) holds true for his literary accounts of Belfast as an account of 'the underworld, the otherworld, the in-between worlds that is not ostensibly the real world.'[21] That liminal status extends to the author's name, itself a linguistic skeleton key to the sectarianism and concomitant semantic ambiguities lodged in 'simple' middle-class Belfast life. Carson addresses the juxtaposition between his Celtic and Catholic-sounding first name, 'Ciaran,' and the paternal name, 'Carson,' 'perceived as the epitome of Protestant nomenclature,' due to the legacy of Edward Carson, the barrister who battled Oscar Wilde's libel suit and who successfully led Ulster opposition to Home Rule in

Belfast.[22] In ruminating about the factional tensions programmed, as it were, into his name, Carson defuses its ancestral Catholic and Protestant genealogies by transforming it into a fable about ritualised ambiguity. Most significantly, the autobiographical tale foreshadows that fabulist-philosophical duality given maximum play in *The Star Factory*:

> On Easter Sunday my paradoxical Catholic father William Carson, or Liam Mac Carráin, as he defined himself in Gaelic mode, would bring us children to the slopes of Stormont, seat of the Northern Irish parliament, where, presided over the by giant statue of Edward Carson, we would roll our eggs. Whether this was a subversive act of one of reconciliation I cannot tell, but its ambiguity mirrored that of the family history.[23]

The semi-rural familial scene ('on the slopes') complete with a pagan-esque egg-rolling ceremony beneath a monument to Unionist hegemony (Stormont, Edward Carson) underscores how the author's name mirrors Belfast's divided cityscape. The parents are humorously implicated, choosing that civic site for a sort of symbolic Easter uprising disguised, in Carson's retelling, as an innocuous familial gathering. Oscillating between such coded languages and referents, *The Star Factory* positions the acts of transliteration and translation as its pervasive narrative responsibility. Translation, Carson's vocational endeavour, is figuratively speaking an operative metaphor for how his memoir discerns Belfast's 'rambling ambiguity,' thereby producing the 'subjunctive mood' of his autobiography, 'tensed to the ifs and buts, the yeas and nays on Belfast's history.'[24]

The foundational doubt about language, and its relevance to writing Belfast, is practised through the verse and prose-poetry in *Belfast Confetti*, as in a passage pursuing the geographical and geological etymologies behind the word 'brick':

> Belfast is built on *sleech*—alluvial or tidal muck—and is built *of* sleech, metamorphoses into brick, the city consuming its source as brickfields themselves were built upon; *sleech* this indeterminate slobbery semi-fluid [...] this gunge, allied to *slick* and *sludge*, slag, *sleek* and *slush*, to the Belfast or Scots *sleekit* that means sneaky, underhand, not-to-be-relied-on, becoming, in the earnest brick, something definite.[25]

The Star Factory restarts this exploratory tactic. Belfast originates in Irish, in English, and in nonverbal language systems like signage, emblems, iconography, artefacts, fashion, maps, directories, packaging, and products.

Largely distanced from interpersonal relationships, the autobiographical narrator develops his identity as a solitary multilingual interpreter or tour guide introducing and reintroducing Belfast to himself and to readers through circumspect translations of the city's physical properties occasioned by Wordsworthian 'spots of time' that flit through personal memory. The work is consequently tangential and non-representational; the clarifying, open-ended discourse around interpretation means that the city surfaces within hypotheticals and measured deductions and inductions teased out by a memoir, performing a semantic séance on behalf of Belfast's superstructures and substructures, past and present. One such exemplary object is a former gardener's cottage, called 'The Bungalow,' in which Carson resided for over 20 years while working for the Northern Ireland Arts Council. The renaming of the building The Bungalow derived in the 1940s from popular local shorthand for its similarity to such houses. In turn, Belfast's utility providers made the nickname into an official address, using it for their official correspondence and billing. Carson's interest in the metamorphism behind the name illustrates the organic unpredictability of language and thus warrants a fable-like etymological digression that typifies the prose style of *The Star Factory*:

> 'Bungalow' in English sounds like a brand of coal or anthracite; more interestingly, if accented on the last syllable, it can be exactly transliterated into Irish as 'bun na gcló', a phrase redolent of ambiguity, but almost impossible to translate. 'Bottom of shapes' is a possible interpretation [...].[26]

Language and names are ceaselessly mobilised, a phenomenon epitomised by the poem 'Belfast Confetti' wherein 'confetti'—referring to stone and glass projectiles—takes, in all at once, concrete material, figural expression, and urbanised gesture, the city's general 'riot squad,' 'raining exclamation marks, / Nuts, bolts, nails, car-key [...] a fusillade of question marks.'[27] Likewise, in writing about post-war, working-class Belfast of his youth, the text focuses on how raw materials attach to any given era's variable idioms and vice versa, recovering anarchical substantiality essential to language that turns bloodless words into corporeal objects and exotic or once-anonymous objects into tangible words, foreclosing the disembodied or incorporeal abstractions on which official, institutional, and administrative cultures derive their authority.

Excavating everyday philological fields loaded with meanings, *The Star Factory* links Belfast's recent past with archaic meanings that reside in objects and words, an anthropological approach catalysed by modelling his prose on the Irish oral tradition. The memoir re-enacts and dramatises how its narrator's younger self studies and replicates the father's inexhaustible oral storytelling skills.

An educated, autodidact postal worker fluent in Irish and the cross-cultural constructed language known as Esperanto, Carson's father, Liam Mac Carráin, functions as an unsuspecting language guide and storytelling mentor for the younger Carson whose nascent desire to narrate his own Belfast memories materialises, in Proustian fashion, within the 'warp and woof' of the 'memory theatre' an intimate familial space suffused by the father's exemplary oral tales.[28] This memory theatre exists on two planes of urban reality. Most obviously, the 'memory theatre' represents the unfolding memoir *The Star Factory*. More significantly, the 'memory theatre' is a privileged social space, mainly between father and son but extending outward to include unnamed fellow Belfast residents who participate in the telling and retelling stories about the site known locally as The Star Factory, a demolished building at 322 Donegall Road that once housed a boys' clothing factory. The apocryphal lore about that site's other historical uses gives communal impetus to kaleidoscopic 'versions' and 'parables' about it, as well as reports about uncanny Belfast incidences said to have happened in and around the former factory's 'Zone,' a psychosocial area described as 'an interactive blueprint; not virtual, but narrative reality.'[29] In this latter iteration of 'The Star Factory,' and its Zone, the unreal-real city locale represents a centrifuge from which Carson's Belfast memoir draws power and authority, as the father's verbalised stories about the building's function (as an infernal salt mine, or dilapidated treasure house, or landing pad for space aliens) conjure alternative city legends and imaginary urban histories ('versions' and 'parables'). The narrator explains that the 'stars' stand for mnemonic devices and 'crucial points' within the 'constellations' and 'patterns of the [urban] everyday' within an oral recitation or urban folktale. Each star-punctuated story is manufactured within that socio-familial space, a real, intimate Belfast 'factory' in the oral tradition, displacing the industrial-era, commercial enterprise that was once the 'real' Star Factory, that disused topographical locus useful now only for building Belfast stories about its presumably colourful past and fantastical interiors and recesses.

Moreover, the father's published volume of Irish-language stories, entitled *Seo, siúd, agus Siúd Eile* (1986) subtends *The Star Factory* as its *Ur*-text. Carson translates the Irish book's title as 'Here, There and There

Again,' 'This, That, and the Other,' and 'Miscellanea' categories befitting the spatially oriented wanderings of *The Star Factory*.[30] Characterised by digressive yet tightly woven patterns, the father's retold myths match the topographical intricacies, equivocal names for objects, and meandering intractability in the concrete jungles of big city Belfast. Remarking on the muse-like status of the father as 'master [oral] storyteller,' the narrator announces that the present autobiographical work about Belfast will emerge in corresponding 'latitudes of anecdote and parable' with 'stories re-invented' through a 'hodgepodge' and 'mish-mash' of 'legend, sagas, folklore and songs.' Claiming an ahistorical scope analogous to that in his father's Irish orality, the memoirist is 'incapable of being here and now without remembering the previous narrative zones he's passed through.'[31] In dispensing with chronology, the memoir's nonlinear layout establishes its extemporal form through mnemonic devices and thematic cues, or pivots, modelled on the father's storytelling and culled from addresses and street names and intersections in the shared familial homes around working-class Belfast. The text foregrounds the father's improvised and accidental pedagogies. Carson's story about Belfast is enmeshed in his listening to his father, and this listening involves a gradual education in how to defy historicism, realism, and naturalism through the timeless discourse and discipline of Irish orality:

> The serial mode [of the father's storytelling] allowed ample scope for such [fantastic] scenarios, whose iconic details might be mirrored over many episodes, in different shifts of emphasis or content. At such points, my father's voice would elevate and quicken, since remembering the narrative depended on these rhythmic clusters or motifs. Compressed mnemonic devices, each contained within itself the implications of past and future.[32]

The memoir opens at '100 Raglan Street' as the Irish-speaking father sits on the 'throne,' smoking in 'the cramped dimensions of an outhouse,' telling the captivated son a mythical tale. In harkening back to that boyhood scene, the distracted narrator limns the memory with allusions to the royal court and to 'chamber pots,' drawing humorous philological associations from the jargon about toilets, plumbing, and pipelines, which lead to the childhood fantasy, among his young peers on the Belfast streets, that 'sewer covers were portals to a parallel sub-universe; embossed with arcane lettering and numerals, [and] their enormous, thick, cast-iron discs.'[33] The philological sequence culminates with the 'library' being redefined as euphemism for 'water closet,' the latter image leading to remembrance of a childhood closet in which he imagines stepping through its 'mercury' into another era, wandering Belfast's predawn streets, and riding a tram's 'iron parallels of time' wherein through a 'gaslit thoroughfare' of that archaic dream city watching real fruit sellers setting up their market.[34] The father's ritualised oratory steers the narrator's imagination and memory through subterranean *Belfasts*; the vernacular for domestic objects—'library,' 'water closet'—propels further projections; the 'real' Belfast of the early 1950s, with its outhouses, sewers, manholes, and libraries, emerges inseparably from the phantasmal city conjured by the son's unfolding memoir, a text sprung from the preliminary spell of the father's timeless voice.

Preserving the *mythos* of Irish storytelling and its embodied voices within a mid-century, post-industrial urban context develops as a meta-textual objective throughout the memoir. 'Radio Ulster' celebrates the family radio's 'big warm hugging Bakelite body' with voices resounding through a favourite BBC Radio Ulster programme named *Tearmann*, 'Irish for sanctuary,' leading to the Irish, '*Ba ghanth liom mé féin a chur i bhfolach innti*,' which the narrator translates into 'I used to hide myself within her.' From that maternal image, the text delves into a recitation of a tale from *Knights of God*, an anthology of ancient stories about saints given to him by his father, with its primordial tales around masculine heroes such as St Finnian and Ciaran.[35] The lives of the saints, transcribed from *Knights of God*, then revitalise the domestic ritual of listening to Radio Ulster a 'disembodied medium' simultaneous to the father's physiognomy and full-bodied eloquence.[36] Carson conflates these pan-historical cross-sections of experience (Radio Ulster, Irish etymology and proverbs, stories of the Catholic saints) with a kind of familial eccentricity and insurgency, highlighted by a story about his father's arrest during the Second World War, when he had been mistaken for 'Uncle Pat [who was] rumoured to have been in the IRA.' Like the waning art of oral storytelling, the Irish Republican Army (IRA) symbolises an Irish hiddenness lurking in the confines of Belfast:

> Outside of such photo-calls, IRA men were practically invisible [during my youth] [...] seeming to exist by rumour or osmosis in a narrative dimension largely inaccessible to the overwhelmingly non-combatant Catholic population. I used to think of them secretly meeting in minuscule cells built into cavity walls, lying parallel in threes beneath the floorboards of a 'safe house'

[…] I was sure they knew the sewer system of the city inside-out, where they were won to flit like wills-o'-the-wisp from manholes, into culverts, into niches where they'd stand like statues as the dark police passed underneath unwittingly.[37]

In another memory illustrating how ancestral storytelling articulates contemporary urban unreality, Carson reiterates an oral Irish tale from his father, transcribing into English, about a blacksmith who falls from the graces of both heaven and hell left to wander the 'moors' wraith-like, taking the form of a 'wandering star,' an image that harkens back to Belfast's White Star Street and the famous British shipping company White Star. In a fugue-like denouement, the father's tale initiates translations about the Irish word *sopog* into 'will-o'-the-wisp,' and 'fairy fire,' leading to English literary asides about Friar's Lanthorn, Ignis Fatuus, and Friar Rush (and Tuck), culminating in accounts of playing Robin Hood in Belfast's 'jungle,' a 'willow scrub-land' on the 'bank of the Owenvarragh river' until his family moves to Mooreland Drive into a home the 'underlying hollow dimensions' suggestive of trap-doors concealing Sherwood's Merry Band, or even 'an IRA man on the run.'[38] The memoir braids disparate threads of the city's unreality and reality—the Belfast father and his oral tale of Will Gallagher, the doomed blacksmith reincarnated as the flame-like 'will-o'-the-wisp,' an image that then metamorphoses into figures from English folklore and literature, captivating the young reader in Belfast who then playacts that literary material within the undeveloped spaces of a working-class neighbourhood, evoking through that recreation both the fugitives in Robin Hood's gang and the outlaw IRA.

Recounting childhood schoolroom rituals in St Mary's Christian Brothers' Grammar School on Barrack Street in Belfast, Carson explicates clandestine games about disappearance. A designated classmate would hoist himself up and out of the schoolroom window, temporarily vanishing onto the ledge outside, or, conversely, climb into a rack of coats in the schoolroom and burrow horizontally and 'foetally' inside them, disappearing from sight. Those audacious acts would win the doers the title of that day's 'Invisible Boy' who was 'simultaneously free and unfree, there and not there; his reality depended on the observer.' This phantasmal play with seeing and not seeing, presence and disappearance spilled into enduring daydreams about being a Mafioso or IRA 'man on the run,' trapped between Catholic-dominated Falls and Protestant-dominated Shankill Roads where he 'desperately searches for a manhole' in which to escape.[39]

The need to escape polarising Protestant and Catholic codes recurs when classmates take a forbidden route home and encounter menacing older boys from the other side of the sectarian division who bully them into choosing between the flag of the Irish Republic and the Union Jack.

In a panoptic city permeated by surveillance even before the onset of The Troubles, verbal interrogation and creative acts of invisibility develop productive, if ominous, relevance.[40] The memoir recounts Catholic school pedagogy predicted on rote learning, 'conjugations and declensions' and memorisation of names and 'hierarchies' with 'liturgical exactitude' along with 'different techniques' and 'subtle disciplines' for the 'universal administration of pain' when the answers are wrong.[41] The anticipation about being beaten with the 'strap' triggers recollection of the 'Manufacturer of Straps for the Discerning Educator,' a real bootmaker with a phantasmal-sounding name located at 68 Divis Street, confirmed by the 1948 Belfast directory, published the year of Carson's birth. This, the banal loggings, in a municipal directory are transmuted into a tale about sadism and powerlessness.

However, not all Belfast memories conform to that rhythmic framework modelled on the Irish oral tradition. Carson's text identifies features in Belfast that defy such narration while still demanding autobiographical scrutiny and memorialising. Often these impervious facets of the city earn the status of an 'aura,' an elusive category of knowledge which recurs throughout the memoir and which the narrator finds especially useful when analysing the 'Zone,' that socio-cultural storytelling space inspired by the name and location of a defunct Belfast clothing manufacturer. 'The Zone was not the Factory,' Carson writes, 'but it was of the Factory and bore its aura.' In deploying this term, derived from theosophy and religious mysticism, Carson knows the redefined conceptual significance of 'aura' is grounded in secular, materialist philosophy and, more importantly, in Modernist theories about the unreality of city life, ideas developed outside of Irish culture and, for that matter, quite apart from the Anglophone world.

Modern Aura: Belfast as Phantasmagoria

Irish traditions of oral storytelling and fable, steeped in voice, punctuated by changing tones and recurring cues, translated and transliterated from Irish into English, and channelling the lore of afterlives, ghosts, wraiths, wanderers, and heroes, prove remarkably effective as autobiographical ciphers through which Carson smuggles memories of a gritty Belfast childhood that was steeped in comparably liminal states and in contemporary, urbanised experiences of invisibility and timelessness. As I have argued, much of that autobiographical transmutation of the archaic into the contemporary and of present urban actualities into antique Irish archetypes relies on Carson's dexterity with Irish and English and on the autobiographer's congenial agnosticism about whether Irish and English languages can capture reality. But even while testing linguistic limits against the concrete features of the unreal city, Carson's *The Star Factory* catalogues uncanny visual memories and aesthetically charged encounters in Belfast that exceed the grasp of such narration. In doing so, the memoir draws heavily on two distinctly Modernist principles about the city—phantasmagoria and aura—interrelated categories of experience and knowledge developed in the 1920s and 1930s, in writings about Berlin and Paris by German intellectual Walter Benjamin, whose frequently invoked presence runs throughout Carson's writing about Belfast.[42]

As a materialist thinker and a Marxist philosopher with a learned background in Judaic mystic traditions, Benjamin theorised about methods that might undo the neutering, or dulling, of personal identity caused by technology-driven consumerist capitalism embodied by the metropolis. In Benjamin's view—one implicitly shared by Carson—a stagnating process of reification conditions its urbanised citizenry to be docile once they conceptualise their existence as a commodity, one more object in the sphere of everyday production and exchange. Benjamin's figure of the urban wanderer, or *flâneur*, stands as the anti-hero opposed to this ossification, rising as an alternative, active method for knowing oneself and one's city, a model taken up by Carson's peripatetic narrator in *The Star Factory*. The *flâneur*—the detached, attentive spectator culled from Charles Baudelaire's urban poetics—pondered the 'religious intoxication of great cities,' a principle echoed by Carson in the autobiographical statement, 'I am in religious awe of the power of [Belfast] names.'[43] Benjamin traced how Baudelaire's intoxicated *flâneur*, a nonconformist and 'illuminati,' whom the poet himself found in Edgar Allan Poe's story 'The Man of the Crowd' (1854), was reinvented by Surrealist novels and in the sensory shocks registered by the meandering narrator in Marcel Proust's introspective epic *In Search of Lost Time* (1913–1927). More than a stereotype or poseur, the *flâneur* represents an alternative form of urban consciousness, a sort of double agent, who exists within the city and yet moves with a level of detachment that provides unique access to the fugitive meanings largely unregistered by the crowd. In imaginative literature as in real life, the *flâneur* saunters about manufactured spaces, neither a machine-like worker nor a sleepwalking consumer. As a detached outsider, he rips off the optical blinders and sensory filters imposed by civic conformity and functional pragmatism. Instead of submitting to the fate of commodified subject or capitalist tool, the *flâneur* wanders the city, studying randomness, changeability, and ephemerality. Carson's narrator re-enacts past wanderings around Belfast in exactly such solitary postures, contemplativeness, and dispassionate observation, turning the autobiographer into an Irish version of this Baudelairean ideal.[44]

Benjamin's works on Berlin and Paris postulate that, however unfathomable cities seem, being technologically and ideologically determined, the cityscape constitutes a magnetic field bristling with meaning. Begun in 1928, and unfinished when he died in 1940, *The Arcades Project* is Benjamin's hydra-headed 'exploration of the soul of the commodity' and, like the Irish oral traditions and the father's storehouse of narrative modalities, subjects, and themes, so too does Benjamin's *Arcades Project* provide a decisive formal model for *The Star Factory*. Culling and transcribing a century's worth of literary texts and writings by a range of thinkers involved in the planning and the construction of modern-day Paris, Benjamin's *Arcades* dissects urban enclosures and apparatuses invented by these modern technicians and utopian architects. Such *techne* include mass-produced clothing, shopping arcades, railroad stations, train cars, automobiles, panoramas, bookstores, kiosks, stalls, movie houses, and large palaces constructed for world's fairs, almost all of which figure in Carson's similarly analytical renditions of modern Belfast. Taking these mechanisms apart and examining their psychosocial effects and textures, *Arcades* tracks how city-based technologies replicate, for any user, the time-suspending spatial configuration in dreams, causing transformations in modern consciousness and social organisation unparalleled in human history. To advance this project, Benjamin developed two concepts which Carson draws upon in his autobiographical investigations of Belfast: the phantasmagoria and the aura.[45]

Phantasmagoria was formerly the general term for panoramas, dioramas, and magic lantern theatres, pre-cinematic methods for optical projection that were technological precursors to moviemaking and virtual electronic forms of entertainment. These original forms of public entertainment, refining the ancient *camera obscura*, took root in late eighteenth-century Germany and spread to Paris, London, and American cities. These popular installations and enclosures often staged séances, replicated far-flung exotic or rural landscapes, and presented recreations of dramatic historical scenes. Largely refined as an aesthetic term and thus slightly removed from its originally Marxist meaning, phantasmagoria, for Benjamin's Paris as for Carson's Belfast, functions also as an authorial vantage point through which to recognise and then record the alternative reality concretised in the city's prefabricated, resonating envelope, its equipment, apparatuses, shells, and encasings crafted by technology and propelled by consumerist culture, within which the urbanite travels and lives, as if inside a waking dream.

Since not all elements in Carson's Belfast are translatable through language (morphological and lexical analyses of Irish and English) or supple enough to be distilled into oral traditions (fairy-tale plotlines or lithe, fable-like capsule narratives), Carson, in relying on Benjamin's *Arcades* rubric, compiles prose-poetic lists, quotations, annotations, and aphoristic commentary about research material to document Belfast (as Benjamin does for Berlin and Paris) in its intrinsic state, as a phantasmagoric place. Belfast's phantasmagoria appears within the memoir's cavernous recesses and shelters, in sitting rooms and front yards, half shaded lots and garbage-strewn riverbanks, shopping arcades and sweet shops, archival photographs and cinema shows, street lighting and public transportation. The former shirt factory crystallises this phenomenon:

> The Star Factory had been long since demolished, but bits of its structure still lay at the back of my mind. Floating through its corridors, ascending its resounding Piranesi iron staircases, or wading through a flooded loading-bay, I realised that for some time I had confused the Factory with other establishments, or other purposes, and its dimensions had expanded. Exterior adjuncts of itself lay scattered on the landscape like relics of a bombed city [...] asbestos-roofed outbuildings on the margins of abandoned airfields or the skull-and-crossbones signs on electricity pylons and perimeter fences.[46]

The passage surveys how variant physical formations (and deformations) in the cityscape haunt the memory and the imagination of the urban dweller, so that the Benjaminian term 'phantasmagoria,' in addition to naming these unspeakable exterior nonhuman realities, also designates a dislocated consciousness as it is altered by those external fabrications which surround and press upon an over-stimulated body. Human bodies become one with Belfast's concurrently interior-and-exterior phantasmal domain, as in this *ekphrasis* on a street paving scene:

> Then the street would be occupied with military-mortar cement mixers and a team of sappers in mismatching uniforms of greasy serge suits, a pair of whom would tamp parallels along the wet concrete street with a plank set on edge, two sets of rocking-horse handles attached to its ends. For a day and a night, or more, the street would be a no-go zone, demarcated by the serial monocular glow of a red bull's eye oil-lamps hooked onto wooden rails between saw-horses, as the aromatic burning oil you sniffed was cut by acrid-smoke from the watchman's brazier that had red holes punched in it.[47]

The scene renders an infrastructure project within a cinematic vocabulary. That diction attests to the enigmatic and transitory animatronics within the seemingly routine. The work site absorbs human bodies whose exertions set in motion industrial machinations and transmute raw materials that literally make and remake Belfast. Furthermore the passage illustrates the juncture between the dispassionate *flâneur's* idiosyncratic attentiveness, personified by the invisible narrator, and the street's infinitesimal concretions and municipal stagecraft, commemorated through poetic descriptors—'team of sappers,' 'greasy serge suits,' 'rocking horse handles,' 'serial monocular glow.' Belfast as a phantasmagoric city appears even in smaller scale tableau. Recalling hours spent in a billiard hall, Carson draws on an 'inward eye' to track how minuscule interior details summon up the fullness of the space's exterior:

> [E]ach stroke [on the snooker table] left a dot of chalk that on the cue-ball, which in turn deposited a microscopic track across the baize or, more especially, against the nap of it. So, one could reconstruct an epic frame from an examination of the empty green arena, until its supervisor, with a wide brush, wiped away the evidence.[48]

In another nonlinear and phantasmagoric renovation of urban routine, the narrator settles into a Belfast train car bound for Balmoral, 'in a compartment that smells of tobacco and autumnal-coloured moiré cut-moquette upholstery.'[49] In studying more closely the myriad of details within the train, the narrator experiences 'momentary dislocation' and then attends the motion of the train as it speeds along the tracks, dissecting each interrelated convolution in the locomotive whirr, such that the excruciations of textual detailing restore to a banal suburban commute its innately fantastical nature:

> The whole elaborate system of junctions, sidings and crossovers is corroborated by the interlinks of rods and levers, wires plumbed into black tubings snaking parallel to the tracks, under intervallic staves of telegraph wires strung out between high poles, as the sleepers below exude oil and creosote, and the heraldic armatures of railway signals click their intermittent semaphores, trying to orchestrate the movements.[50]

Phantasmagoria colours the recollections about Belfast movie-going as well. In diction that directly borrows phraseology from Louis Aragon's *Paris Peasant* and Walter Benjamin's *Berlin Childhood*, Carson's narrator recounts a 'palpable dream' of being 'pleasurably lost' in a Belfast 'precinct crammed with shops, stores, offices, public houses, cafes cinemas' including a semi-fantastic 'sleep department of a vast emporium.'[51] Suffused with dream content, the autobiographical vectors about the area housing the 'Ulster Cinematograph Theatres' are nevertheless anchored through actual coordinates, the 'dense urban space of Arthur Street and its confluence of five streets' and a 'quadrangle of small shops containing three covered arcades of junk stalls, mostly built in 1848.'[52] Alluding to childhood screenings of films that depict competing parallel universes—*The Wizard of Oz* and *20,000 Leagues Under the Sea*—Carson explores how childhood fantasies drawn from films develop into adult imaginings as well, especially in autobiographical correspondences found in British director Carol Reed's noir thriller *Odd Man Out*, based on the F.L. Green novel. Reed's film, profoundly invested in Belfast's interior-exterior enclosures, recesses, and fissures, serves as a cinematic exemplification, par excellence, of Walter Benjamin's theories about the enervating potential in moviemaking.[53] Thus in Carson's memoir, Reed's film provides a cinematic corollary and a reflexive text authenticating *The Star Factory*'s temporally disjunctive wandering within his version of Belfast. Shot almost entirely in the post-war Belfast of Carson's youth, Reed's film follows the street-bound odyssey of Johnny McQueen (played by James Mason). McQueen, a mortally wounded leader of a local IRA unit being pursued by the police, hides in pockets throughout the city while encountering a cross-section of Belfast citizenry who interpret him in various degrees of ambivalence and captivation, as if McQueen were an urbanised extension of the city's troubling inscrutability. Through memoires about seeing Reed's film in Belfast, Carson obliquely reveals how the fragmented autobiographical tableaux in *The Star Factory* mirror the confined dislocations in Belfast undergone by Reed's doomed hero. Although Reed's film opens in realistic mode with crooked aerial shots of Belfast Lough, McQueen's disorienting separation from fellow revolutionaries relatively early in the film suspends the advancement of linear time, opening out to a non-chronological, collage-like film which Carson's memoir consciously imitates. Removed from social relationships that narrowly defined him, McQueen undergoes what Carson, alluding to his often solitary Belfast childhood wanderings, terms an 'internal exile' within Belfast, underscored by the film's moody 'Doctor Caligari camera angles.'[54] This Reed's cinematographic style provides a generative model for Carson's elliptical literary montages; both texts, film and memoir, blur the normal boundaries between public and private, outside and inside. In Reed's film such spaces include the ramshackle lodging in a safe house, a bourgeois sitting room, spacious pubs and saloons, a dance hall, a church rectory, an artist garret, a junkyard, a bomb shelter, a getaway car, a tramcar, horse-drawn hansoms and automotive taxis, a dockland and innumerable storefronts, alcoves, lanes, alleys, and street corners.

Analysing the phantasmagoric paradox of interiors that double as exteriors, Carson decodes Reed's Belfast in terms that apply equally to his version of the city, in that Belfast, '[in *Odd Man Out*] is a Daedalian construct in which even the street scenes, with their strong Caravaggio chiaroscuros, look like interiors.'[55] Compatible to Benjamin's theory that commodity capitalism alters urban experience such that the cityscape conjures an irrealism collapsing distinctions between waking consciousness and dreams, the narrator of *The Star Factory*, like James Mason's character in *Odd Man Out*, is unable to distinguish dreams about Belfast from actualities in the city. Both Mason and Carson's narrator experience acute sensory flashbacks about encounters in particular Belfast locations that conjoin disparate past experiences, forming these into singular, hallucinatory scenes. Such moments document how the individual subconscious reshapes urban reality through its own sublimated imagery. *The Star Factory* adapts this montage technique to its own ends. The text deliberately 'cuts away' from descriptions of 'real life' past happenings to narrate scenes from recurring dreams. The memoir describes a childhood dream about constantly missing a particular red bus at Dunville Park and also around Ballysillan, anachronistically set in 1930 'when there were no buses' and a dream vision of St. Peter's square that has been reconfigured 'with a piazza' and buildings modelled on various cities, 'Belfast grocer's corner shop, Parisian boutique, New York diner, Dublin pub, Warsaw synagogue, Berlin brothel.'[56]

Ultimately, Carson envisions Belfast's concretions, both within and without the memoir *The Star Factory*, as emanating a timeless ineffability corresponding to what Benjamin describes as the reflexive 'aura,' affecting seer and seen alike and having its most recognisable nonhuman manifestation in commonplace objects and in artworks, including photography and film. Benjamin defines the aura as 'an ornamental halo in which the object or being is encased as in a case,' and, reflexively, as 'the distance of the gaze that awakens in the object looked at.' In a technological metaphor for how the human consciousness seizes on an aura within its attended appearance, Benjamin describes the aura as being 'developed in the darkroom of the lived moment.' Acknowledging the opacity of the ordinary world is a prerequisite for recognising and realising the aura of things. According to Benjamin, '[w]e penetrate the mystery [of the aura] only to the degree that we recognise it in the everyday world, by virtue of a dialectical optic that perceives the everyday as impenetrable, the impenetrable as everyday.'[57]

Carson's memoir confronts and joins the formal and autobiographical challenges posed by Benjamin's theory of the aura, mainly through the memoir's curated contemplations of discrete objects and scenes. Examining engravings and photographs from frozen moments in Belfast's ordinary past in 'The Panoramic Photograph Company,' the chapter unlocks a fleetingness within the seemingly arrested scenes of Belfast-past. An engraving from 1786 depicts a scene set on High Street, offering the narrating witness an opportunity to study the impenetrability of the small grouping of people and to unleash the fugitive haze, or aura, contained within that graphic transience. The autobiographical moment, charged with creative visioning of an art-object, serves, as do myriad moments in *The Star Factory*, as a case study in practising Benjamin's 'dialectical optic':

> A tricorn-hatted gentleman is chatting up two ladies, one of whom is looking over her shoulder, possibly at the woman carrying a baby. Two cloaked figures conduct some business, and there are other knots of twos or threes in the background. In the far distance, a gang of men or sailors converge on the tangle of masts, spars and shrouds where the Farset debauches into the Lagan.

The aura gleaned from the High Street engraving is realised (made real) within the medium of the hyper-perceptive text itself. The uncovered spontaneity dwelling within the image of a long-gone moment in late eighteenth-century Belfast finds corresponding, ***dialectical***, vitality in the late twentieth-century words of an autobiographer who responsively intuits the figures' motives and moods, fulfilling Benjamin's mandate that the seer's receptiveness to the object will yield up 'distance of the gaze' awakened in the object looked upon—its aura. This means the aura is both a means and an end in Carson's writing. As Benjamin scholar Hansen notes, 'aura is not an inherent property of person or objects but pertains to the medium of perception, naming a particular structure of vision' and 'a phenomenal structure that enables the manifestation of the gaze, inevitably refracted and disjunctive and shapes its potential meanings.'[58]

When the narrator studies his stamp collection—at first glance the most mind-numbing autobiographical topic imaginable—aura again is made manifest. The text actualises the aura within the postal relic and in doing so transcends the banal, propagandist historicism stamped on the object by Irish national bureaucracy. The stamp in question, a 1941 issue known as 'The Gunman,' commemorates the 25th anniversary of the 1916 Easter Rising in Dublin. Carson acknowledges the Nationalist import of the relic, its 'subversion of authority [...] for nowhere was the Crown as near ubiquitous than as on postage stamps, these little emblems of the temporal realm.'[59] Yet that static historicism means little; the third-rate portrait of the gunman is 'banal, pious, badly drawn,' and, as an item for potential sale, the stamp is 'next to worthless in monetary value.'[60] Its insipidity and ugliness nevertheless contains within it an aura that 'fascinates':

> I love the blue-black ink that seems to have a tint of bottle-green in it, so that it summons the dull enamelled frames of Royal Constabulary bicycles [...] the colour of gunpowder, broken slates or magnets; the ooze blue-clay of the Lagan at low tide; coke-smoke from the gas-works; livid live lobster blue; rubber bullets, purple cobblestones, a smear of blackberries; cinder-paths at dusk, when no one walks on them; the black arm-band of the temporary postal worker.[61]

From smudged colours closely examined through a magnifying glass trained on bureaucratic detritus, *The Star Factory* cultivates language and subject, Belfast, as aura. The city's unreality stands equal to the reflexive, mesmeric text. And the written word unearths legible opacities from Belfast's impenetrability.

Notes

1. Ciaran Carson. *The Star Factory*. New York: Arcade Publishing, 1997, p. 48.
2. Ibid., p. 122.
3. Ibid., p. 133.
4. *The Star Factory*, p. 126.
5. Denis Donoghue. *Being Modern Together*. Scholars Press, 1991, p. 12.
6. T.S. Eliot. *The Waste Land: A Facsimile and Transcript of the Original Drafts Including the Annotations of Ezra Pound (A Harvest Special)*, edited by Valerie Eliot. Harcourt, Brace Jovanovich: New York, 1974, p. 136. In addition to describing London crowds as 'unreal,' Eliot describes the city as a site of 'unreal emotions and real appetites.'
7. Italo Calvino. *Invisible Cities*. Harcourt: New York, p. 44.
8. 'A Life in Poetry: Ciaran Carson' Interview with Aida Edemarium. *The Guardian*. 16 January 2009.
9. Ciaran Carson. *Exchange Place*. Flagstaff Press: Belfast, 2012, p. 94.
10. Ciaran Carson. *Belfast Confetti*. Salem: Wake Forest University Press, p. 67.
11. Ibid., p. 35.
12. Ibid., p. 58.
13. In part three of this chapter, I examine Carson's deep intellectual and aesthetic investment in the writings and thought of the German philosopher and urban social theorist Walter Benjamin (1892–1940), who elaborated theories about late capitalism and the *flâneur*, *phantasmagoria*, and *aura* in *One-Way Street* (1928), *The Arcades Project* (1928–1940), and, most famously, in the essay on photography and film, 'The Work of Art in the Age of Mechanical Reproduction' (1938). In a July 29, 2017, email, Carson informed me that Walter Benjamin 'is a presence' in Carson's novel *Exchange Place*, a Belfast story which Carson describes to me as a 'fictional counterpoint to The Star Factory.' Prose extracts from Benjamin's *Berlin Childhood* (1928) form the opening epitaph in Carson's poetry collection *Belfast Confetti* (1989), and Carson's novel *Exchange Place* (2012) uses Benjamin's concepts for certain chapter titles and the novel quotes many passages from Benjamin's writings. I argue that in composing *The Star Factory*, Carson models the intertwining of municipal histories and etymological, genealogical, and morphological analyses within a deeply personal memoir on the blueprint for that still largely taboo hybridity inaugurated by Benjamin's incomplete but gargantuan *Arcades Project*.
14. Alexander, Neal. *Ciaran Carson: Space, Place, Writing*. Liverpool: Liverpool University Press, 2010, p. 62.
15. *Belfast Confetti*, p. 35.

16. 'A Life in Poetry: Ciaran Carson' Interview with Aida Edemarium. *The Guardian.* 16 January 2009.
17. Seamus Deane. *Strange Country: Modernity and Nationhood in Irish Writing Since 1790.* Oxford: Clarendon Press, 1997, p. 96.
18. Seamus Deane. *Strange Country*, p. 97.
19. Ciaran Carson, Interview. *Chicago Review* 45:3 (2000), p. 96.
20. Ibid., p. 97.
21. Ibid., p. 99.
22. Ciaran Carson. *Last Night's Fun: A Book About Traditional Irish Music.* New York: North Point Press, 1998, p. 181. Despite its light-hearted title and characteristic comic flourishes, Carson's autobiographical study of traditional Irish music is a rigorous exploration of how the Irish language intersects with the culture of music and dance and how the structural components unique to traditional Irish music have corollaries in Irish and even non-Irish literary forms.
23. Ibid., p. 181.
24. *The Star Factory*, 76, and *Belfast Confetti*, 67. By 'vocational endeavor' I refer to Carson's professional work as a literary translator of Italian, French, and Irish classical and modern literatures, respectively: *The Inferno of Dante Alighieri* (Granta Books, 2002), *The Alexandrine Plan* (Wake Forest University Press, 1998), *Midnight Court* (Wake Forest University Press, 2005), and *The Táin* (Penguin Classics, 2007).
25. *Belfast Confetti*, p. 72.
26. *The Star Factory*, p. 52.
27. Ciaran Carson. 'Belfast Confetti.' *Collected Poems.* Salem: Wake Forest University Press, 2008.
28. *The Star Factory*, pp. 66–67.
29. Ibid., p. 66, 63.
30. *The Star Factory*, p. 76.
31. Ibid., p. 77.
32. *The Star Factory*, p. 66.
33. Ibid., p. 5.
34. Ibid., p. 4.
35. Ibid., p. 106.
36. Ibid.
37. *The Star Factory*, p. 117.
38. Ibid., p. 151.
39. Ibid., p. 210.
40. For a poignant, harrowing account of Carson's experiences being interrogated by Belfast police and paramilitaries, see the autobiographical prose vignette 'Question Time' in *Belfast Confetti*, pp. 57–63.
41. *The Star Factory*, pp. 212–213.

42. The epitaph for Carson's *Belfast Confetti*, on p. 14, is an extract about getting lost in one's own city and its relevance to writing, from Walter Benjamin's *A Berlin Childhood Around the Turn of the Century*. On page 26 of *The Star Factory*, Benjamin's *Selected Writings Vol 1: 1913–1916* appears in a list of Modernist texts purchased by the author at a Waterstone's shop on Royal Avenue in Belfast. Carson's *Exchange Place* contains too many references to Benjamin and his concepts to enumerate and annotate here. Indeed, critical work remains to be done on *Exchange Place*, particularly its thematised dialogue between Belfast and Paris and Irish letters and French literature.
43. *The Star Factory*, p. 43.
44. For a cutting-edge examination of how Carson's literary wandering in his poetry stands in opposition to Thatcherite neoliberal redevelopment in Belfast, see John Goodby's 'Walking in the City: Space, Narrative, Surveillance.' *Ciaran Carson Critical Essays*. Edited by Elmer Kennedy-Andrews, Dublin: Four Court Press, 2009.
45. In *The Star Factory*, Carson alludes to Benjamin's influence through frequent use of W.B.'s concept of 'aura' and in mentions of Belfast's shopping arcades, as in the closing sentence of the chapter 'Barrack Street II': 'we took French leave and exited into the city, to explore its yawning avenues and dark arcades.' p. 220.
46. *The Star Factory*, p. 61.
47. Ibid., p. 242.
48. *The Star Factory*, p. 80.
49. Ibid., p. 84.
50. Ibid., p. 242.
51. *The Star Factory*, p. 222.
52. Ibid.
53. In addition to its prevalence in *The Star Factory*, Reed's film *Odd Man Out* is examined at length in a chapter entitled 'Odd Man Out' in *Exchange Place*, pp. 56–60. For the interconnection between cities, phantasmagoria, and modern filmmaking, Walter Benjamin's 'The Work of Art in the Age of Mechanical Reproduction.' *Illuminations*. New York: Schocken Books, 2007.
54. *The Star Factory*, p. 131.
55. Ibid.
56. *The Star Factory*, p. 197, 199.
57. All above quotes from Walter Benjamin's definitions of aura have been transcribed from Miriam Bratu Hansen's invaluable essay and collations of W.B. texts on this long-debated concept of the 'aura,' 'Benjamin's Aura' in *Critical Inquiry*, 2007, pp. 339–330, 342, 358.

58. Miriam Bratu Hansen 'Benjamin's Aura,' p. 342.
59. *The Star Factory*, p. 35.
60. Ibid.
61. Ibid., pp. 35–36.

References

Alexander, Neal. *Ciaran Carson: Space, Place, Writing*. Liverpool: Liverpool University Press, 2010.

Aragon, Louis. *Paris Peasant*. San Francisco: Exact Change, Boston, 2004.

Benjamin, Walter. *The Arcades Project*. Boston: Harvard/Belknap, 1999.

———. 'The Work of Art in the Age of Mechanical Reproduction.' In *Illuminations*. New York: Schocken Books, 2007.

———. *One-Way Street*. Boston Harvard Belknap, 2016.

Calvino, Italo. *Invisible Cities*. Translated by William Weaver. New York: Harcourt, 1978.

Carson, Ciaran. *Belfast Confetti*. Winston Salem: Wake Forest University Press, 1989.

———. *The Star Factory*. London: Granta Books, 1997.

———. *Last Night's Fun: A Book About Traditional Irish Music*. New York: North Point Press, 1998.

———. *Exchange Place*. Belfast: Blackstaff Press, 2012.

Cohen, Margaret. *Profane Illumination: Walter Benjamin and the Paris of Surrealist Revolution*. Berkley: University of California Press, 1995.

Deane, Seamus. *Strange Country: Modernity and Nationhood in Irish Writing Since 1790*. Oxford: Clarendon Press, 1997.

Donoghue, Denis. *Being Modern Together*. Scholars Press, 1991.

Eliot, T.S. 'The Wasteland.' In *Collected Poems*. London: Faber & Faber, 2017.

Goodby, John. '"Walking in the city": Space, Narrative and Surveillance in *The Irish for No* and *Belfast Confetti*.' In *Ciaran Carson Critical Essays*, ed. Elmer Kennedy-Andrews. Dublin: Four Court Press, 2009.

Hansen, Miriam Bratu. 'Benjamin's Aura.' *Critical Inquiry*, 2007.

Odd Man Out. DVD. Directed by Carol Reed, 1947: Criterion Collection, 2015.

PART II

Disturbing Phantasies and the Uncanny City

CHAPTER 6

'Neither This nor That': The Decentred Textual City in *Ulysses*

Quyen Nguyen

Modernist novels are often preoccupied with urban life and make city spaces their main backdrops. Naturally, as a city represented with the utmost care in the *magnum opus* of the dominant figure of literary modernism, James Joyce's urban space has been a preoccupation of critics of all persuasions: cartographical, psychoanalytic, sociological, and postcolonial. However, Joyce scholarship dealing with the portrayal of Dublin in *Ulysses* is mostly concerned with factuality in Dublin, and 'real' places have too often been considered the single setting for the wanderings of the characters. Joyce's montage of multiple 'Dublins' through a range of historical juxtapositions and varied styles has taken a back seat in Joyce studies. This chapter addresses that neglected underlying cityscape of the novel and proposes a reading of one of the many faces of Dublin in the eighth episode of *Ulysses*, 'Lestrygonians'. With a depoliticised approach, it will show how the city is portrayed from the point of view of the rambling Bloom, a practical Dublin user, and discuss how the narrator offers the surplus of place names, creating the effect of reality, ushering the reader into the real world more intensely than any novel.

Q. Nguyen (✉)
Division of English, School of Humanities, Nanyang Technological University, Singapore, Singapore
e-mail: thihanhq001@ntu.edu.sg

M. Beville, D. Flynn (eds.), *Irish Urban Fictions*, Literary Urban Studies, https://doi.org/10.1007/978-3-319-98322-6_6

With a deficiency of description, *Ulysses* constructs an unintelligible world, enfeebling the imagination of the reader and thus problematising the production of meaning. This effect is enhanced by repetitive and distancing style that depicts Bloom's walking patterns and denies the reader any chance of imaginatively re-enacting Dublin. The real and modern city of Dublin inundates Bloom's mind with stimuli, conflating the past and present with various locations, bringing forth different figurative and literary images of the city, thus subverting not only the notion of a realistic city but also of wholeness in time and space. The city thus transforms into a signifying field in which proper names, and what they trigger, become signifiers and signifieds, rendering reality multidimensional. The chapter will demonstrate how Joyce's Dublin exists between the interplay of two discourses, as neither a realistic nor figurative city, but rather a textual urban space rejecting totalisation.

'We Are the Real': Factual Dublin and Place Names

In a letter to Grant Richards on 5 May, 1906, Joyce explains his purpose of writing *Dubliners* as 'a chapter in [...] moral history' because the city is a centre of paralysis. Acknowledging his employment of a stylistic 'scrupulous meanness',[1] Joyce vows to be realistic: he will serve what he believes, what it calls itself the truth. Facts were a preoccupation for Joyce when he incessantly queried his brother and his Aunt Josephine about Dublin's minute details in the process of writing *Ulysses*:

> Is it possible for an ordinary person to climb over the area railings of no 7 Eccles street, either from the path or the steps, lower himself from the lowest part of the railings till his feet are within 2 feet or 3 of the ground and drop unhurt.[2]

With a style more intense than in *Dubliners*, Joyce inundates this novel with raw materials, including real names of places and people from *Thom's Directory*. Ian Gunn et al. in their extensive and well-researched publication on topographical Dublin compile a list of all the addresses that Joyce uses from *Thom*'s which comes to a total of 541. While *Oliver Twist* by Charles Dickens, one of the greatest realists and city writers, with London as its basic setting, has only 193 place names in total,[3] the proper names used in *Ulysses* are almost triple those of Dickens' novel. My examination of the number of Dublin place names used in Bloom's episodes unveils Joyce's writing method: in 'Calypso', he uses nine place names, 'Lotus Eaters' 20, 'Hades' 34, 'Lestrygonians' 73. The sheer numbers speak volumes. The density of place names makes the reader pay full attention to the fact that he is reading a text fully attached to the city of Dublin.

Jonathan Culler insightfully notes that, despite the lack of meaning, proper names do have a signifying function, 'they signify "we are the real"'.[4] The place names in *Ulysses* gesture to the objectivity that realism strives to attain. Barthes in *S/Z* provides a name for the excessive details that create this effect of reality: cultural and referential code.[5] Besides four other codes that the reader must decipher when reading a novel, this term designates any element in a narrative that refers 'to a science or a body of knowledge': the cultural code gestures the reader to 'the way the world works', to the type of knowledge that we categorise as historical, literary, physical, and geographical.[6] This referential code works perfectly in Joyce's text to point to the topographical knowledge of the city of Dublin. It informs the reader, not only of places, notable or trivial, in Dublin, but also the city's routes, districts, and landmarks.

Philippe Hamon shares Barthes' opinion about proper names, contending that geographical proper names denote stable semantic entities. Place names in *Ulysses* function like academic quotations and are an economically descriptive tool. They 'guarantee mooring points, re-establish the performance [...] of the referential statement' by coupling the textual Dublin with the extratextual Dublin.[7] This is how Joyce secures the verisimilitude of Dublin: by infusing the text with proper names, he produces 'an overall effect of reality which even transcends any decoding of detail, an effect of reality often accentuated in topographical descriptions'.[8] Due to identical place names, such as Dublin in the literary work and Dublin the capital of Ireland, as with London, Paris, New York, and Saint Petersburg, the reader takes for granted that they are the same. They refuse to accept the thought that *Ulysses* with its 541 Dublin place names might be set in another Dublin. These proper names point from the space where one reads to the space where one knows. Consequently, literary Dublin folds over the real Dublin as though the two were the same.

Furthermore, the reason most critics mix the fictional world with the real world is because the text of *Ulysses* provides no signals at all to discourage the reader from conflating them. Benjamin Hrushovski in his study about the internal and external references in fictions calls the real world the

'External Field of Reference', while the fictive world is the 'Internal Field of Reference'. He suggests that in the text there is always at least one signpost of fictionality 'to separate the text from a description of the External FR':

> though the month and the hour are mentioned precisely, one specific indicator, the day, is floated. This device indicates that the fictional time and space, however closely located in relation to the real world, is somehow suspended above it, has its own, 'floating' coordinates.[9]

The omitted signal may take various forms, for instance, to follow Hrushovski's suggestion, *Ulysses* could have such floating pointers as 'on one June morning of 1904', 'in the year 190*', 'the city D'. *Ulysses* at the very beginning anchors the text to a specific place: the author never passes any opportunity to strengthen the link between his Dublin and the real Dublin by accumulating as many Dublin landmarks as possible: post offices, pubs, stations, streets, and churches. However, the more place names of Dublin are added to the text, the more the reader gets the impression that this might not be a fictional world, and the more specific the time is offered, the more playful the text becomes: this is not the fictional world, not a single signal is available to prove so. With the employment of the place names and the density of these information, the referentiality of signposts becomes the utmost priority in *Ulysses*.

In an attempt to rebuild Dublin, Joyce brings the absent city into presence through place names in which each place name could successfully replace its equivalent place. As Martin Heidegger notes in *What Is called Thinking?* that 'Naming consists by nature in the real calling, in the call to come, in a commending and a command'.[10] Dublin is thus commanded to emerge, to appear, to take shape in the forms of place names. Through this calling, the text supports the naïve assumption that words can stand for objects and language is a purely transparent and satisfactory system of references. Joyce brings real Dublin objects into his text with a determination that his system of place names will not face any semantic problems, and real objects will be locatable and reproducible in text. Nevertheless, as Terry Eagleton observes in *The English Novel*, '[t]o call something "realist" is to confess that it is not the real thing'.[11] Proper names in *Ulysses* have a double bind inherent in them because of their condition of possibility: they are unique and repeatable; they are specific and uncertain; they are signifiers and signifieds; they are real and fictional: they negate themselves and recreate themselves in Joyce's writing. Their accumulation is the key to the uncertainty in the novel.

'Streets Are Named but Never Described': Surplus of Facts, Deficiency of Descriptions, and Failed Production of Meaning

In *Ulysses*, Joyce abandons the intelligible Dublin and goes beyond this norm to challenge the reader's ability to make sense of the city. The act of violating these narrative conventions of describing places starts as early as 'Calypso' and repeats continuously in 'Lotus Eaters' and 'Lestrygonians', when Joyce fills the routes of Bloom with a wealth of real localities to inform the reader about his movement and yet omits all descriptions about the city. Place names in *Ulysses* do not possess evocative power, except as flags that signal to the reader that they are the real. Rather, they become details, a special kind of detail that is inauthentic and hollow because the fictive Dublin is decorated with the spillage of proper names without their imaginative descriptions. Frank Budgen rightly observes how Dublin's places are provided 'without explanations or introductions'. 'Streets are named but never described. [...] Bridges over the Liffey are crossed and re-crossed, named and that is all'.[12] While details play the role of stimulating the reader to construct and develop mental pictures, Joyce hardly instructs the reader on this process. In the 'Lotus Eaters' passage cited above, the whole route that Bloom takes lacks descriptions and is nothing but data, albeit abundant data of place names and details. It is not surprising that Elaine Scarry's study of how literature helps readers imagine and dream via its language absolutely ignores James Joyce and there is not a single image or scene that could make it into her examination. The vividness of objects and places in Joyce are never allowed because he denies the 'imaginary vivacity [which] comes about by reproducing the deep structure of perception'.[13] The only instructions that the reader receives are the movements of the character in the city. Although Dublin is reflected through Bloom's walking steps and his eyes, the unembellished place names enfeeble the reader's power of imagination.

Place names in *Ulysses* are not merely decorations on the surface: they are functional in a more complex way in the process of reading. It should be noted that the Joyce industry generally treats Dublin as a unique one-dimensional object, in which the real places mentioned only serve as nails for scholars to hang their analyses of *Ulysses*-Dublin's historical accuracy on. This approach, as I argue in my '"Dublin what place was it": Making sense of the textual city in Ulysses', strengthens the assumption that the fictive and geographical Dublins are wholly overlapped and secures the place of Joyce's Dublin as seen in Joycean scholarship as the epitome of mimetic writing.[14] Even in all the various attempts of scholars to probe symbolic and historical meanings of Dublin toponyms, this assumption persists: for example, the Martello tower in David Spurr's symbolic reading of the monumental space in *Ulysses* becomes 'the juxtaposition of two archaic forms, one architectural, the other liturgical'[15]; the Nelson Pillar, according to Andrew Thacker's postcolonial reading, is considered 'a monument conceived to signify British imperial rule in Ireland',[16] to name a few, this kind of treatment is still a product of reading that fails to go beyond the realist approach. This tendency is mired in a tradition of reading that tends to go no further than realist interpretations. And yet, I would like to advocate for a depoliticised approach that regards place names in *Ulysses*, despite all their richness in historical and political symbolism, not as symbols, in the sense that they yield no specific meanings useful to the recuperation of the reader.[17] Place names, apart from the signification that they are the real, do not provide any more clues about reality; they do not yield themselves to become material for us to understand Bloom's predilections. In realist novels in which the world is presented 'fundamentally intelligible', Culler frequently reminds us that there are 'causal theories of the relationship between environment or appearance and personality or atmosphere', whence the text of *Ulysses* simply makes a fundamentally unintelligible world.[18] Throughout the novel, the reader never gets any clue to help him at least imagine how the house at 7 Eccles Street is, let alone deducing from this habitat a single characteristic of Bloom's personality. Similarly, the routes that Bloom takes in 'Lotus Eaters' or 'Lestrygonians' yield us no single understanding of Dublin. The text of *Ulysses* goes beyond the conventions of nineteenth-century realism, rich in preparing visual effects, yet does not attempt to justify irrelevant and unnecessary details; and it flaunts its knowledge while still denying the register of appearances. The text, with the surplus of facts, the underprovided details, and the denial of symbolic interpretations, denies the reader any chance of imaginative enactment of an intelligible world of Dublin.

The city in Bloom's episodes is represented from a viewpoint in motion, not from static eyes but in those that walk. It could be said that Joyce provides the stage directions or even that he directs Bloom how to walk, where to walk, and how fast to walk and yet refuses to provide any details about the city. Most omniscient narration in 'Lestrygonians' is written in a very basic style: the sentences are often short and flat, the meanings are clear and straightforward, and the vocabulary is simple and accessible.

He passed the Irish Times.[19]

He stood at Fleet street crossing.[20]

He walked on past Bolton's Westmoreland house.[21]

He went on by la maison Claire.[22]

He entered Davy Byrne's.[23]

The syntax is tremendously simple: subject + verb + prepositional phrase. Bloom walks around Dublin, and the reader finds numerous examples such as 'he walked', 'he passed', 'he crossed', 'he turned', and so on. These bare sentences listing the facts of Dublin work like stage directions and their effects in building a system of instructions for the reader which indicates movements, positions, or attitudes of Bloom. As pointed out by Erwin Ray Steinberg, 'Lestrygonians' shows a strong pattern of omniscient authorial sentences appearing at the beginning of paragraphs or as full paragraphs, operating much like stage directions in a play, to orient the reader.[24] The third-person narrative becomes an impartial camera: it offers no description of the feelings of the character; it strips off adjectives and adverbs; it just captures the movement of the character and the signposts on the way. It also employs a very regular and monotonous style, with legalistic precision.

In her argument about how the mind receives instructions to build the givenness of an object, Scarry hypothesises that it is the 'mimesis of givenness' through the quality of overt instruction in prose that helps change from daydreaming to 'vivid image-making'.[25] Under Scarry's direction, the mind of the reader is asked to follow these instructions in the walking sentences of Bloom:

(Look closely at the walker's legs) His slow feet walked him riverward, reading.[26]

Mr Bloom, *(hear the sounds of his breathing)* quickbreathing, *(look closely at the walker's legs)* slowlier walking passed *(hear the name)* Adam court.[27]

He passed, *(look closely at the way of the characters's walk)* dallying, the windows of *(hear the name)* Brown Thomas, silk mercers.[28]

Consequently, what is conjured up is not the 'givenness' of the places that Bloom passes but rather of his walking movements. In most of the cases, the bare grammatical structure of Bloom's walking movements occupies most of the third-person narration, creating an effect of distanciation and defamiliarisation in Bloom's episodes. Bloom is presented from a distance, and the impassivity of the walking sentences renders Bloom a mechanised object, a chessman on the chessboard of Dublin. The narration is narrowed down to the street level and exactly states the places Bloom passes by. Dublin is immediately reported, as if the character were wearing a Global Positioning Device informing the reader in real time where Bloom is. This style persists throughout the narrative, creating a heightened effect of immediacy and yet strangeness. Although the style functions perfectly in depicting Bloom's movements in Dublin with detachment, this very aim betrays itself with the narrative inundated with monotonous and simple sentences like: 'Mr Bloom moved forward', 'He walked along', 'He crossed under', 'He gazed after'. What kind of literature employs the most basic sentences worthy of beginner-level students of English? These simplistic sentences reveal how the narrative 'parodies its own ability to tell a story'.[29] They are repeated as if being automatically produced. The slight irony is evident in the bare prose of 'Lestrygonians', and the text undoubtedly quotes itself by repeating the bare grammatical structures as mentioned above. The reader constantly encounters 'he walked, he came, he crossed', as if churned out robotically from a writing machine. These sentences alert the reader to the author's intention. They give out a warning as to how they might belong to a complex consciousness, and not simply be identified as the work of a third-person narrator.

With the accumulation of place names taken to extreme levels, the narrative sends out a signal to the reader that it is not only realism but hyperrealism in the way the text aims directly and straight to the geographical Dublin. The distancing style works in a very similar way, mirroring the walking patterns of Bloom, in Dublin's places, evoked specifically by place names. They aim to be a competent medium of representation in which there is no gap between the reference and referent. However, the proper names hollow themselves out, becoming empty signifiers, and the narrative becomes the vehicle of them, alerting the reader to the parody of it. This goes hand in hand with the distancing style, portraying reality on one hand but, on the other hand, ironically repeating its textual structure and highlighting its defamiliarisation.[30] The text imitates hyperrealistic style and yet undermines it at the very time of its employment.

Dublin as a Cobweb of Stimuli and a Semantic Field of Conflated Places

In his essay 'Semiology and Urbanism', Barthes observes how the city is a discourse and a language: '[t]he city speaks to its inhabitants, we speak our city, the city where we are, simply by inhabiting it, by traversing it, by looking at it'.[31] I suggest a reading of Dublin in *Ulysses* as a discourse between the city and its 'user', Leopold Bloom. The city is the writing and the citizen its reader. Each reading is unique and different to another reading, not only in the sense that everyone is an individual but also that each route the citizen takes renders a different reading. The citizen could become a reader of an *avant-garde* work such as Raymond Queneau's *100,000 Million Poems*, creatively producing different verses and meanings, while moving in a city.[32]

Leopold Bloom's odyssey on 16 June, from 8 A.M. at home, through a long day in the streets and private spaces of Dublin, then back to his dark bed far past midnight, offers not only ample opportunity for him to interact with the outer reality but also a reconsideration of his distant past and vivid internal present. Neither comprehensive nor panoramic views of Dublin are offered in *Ulysses*. The narrator with the utmost detachment through the distancing monotonous style intentionally leaves Bloom alone to navigate and interact with the city. Dublin speaks to Bloom via its cityscape with a conviction that he, a Dubliner, can constantly and freely interpret and generate Dublin from his personal and collective memories, from local knowledge as well as popular culture. The everyday life in Dublin is turned into the trivial inventions of Bloom: that is, his obsession with his wife's infidelity, his paltry daily life activities, his failed job, and practical science knowledge, to name a few.

On a deeper level, Dublin has dramatically changed from the very moment Joyce introduces interior monologue: readers unwittingly enter and re-enter the fictive flow of a Heraclitan river-of-reality—a Protean, shape-shifting Dublin or, in other words, an atlas-worth of Dublins. Joyce incorporates various raw materials from the unlimited resources of city life into Bloom's consciousness, creating a sea of sensations and impressions. In a nutshell, the interior monologue narrative mode is an important one of stylistic devices transforming Dublin: it triggers the internalising process in which Dublin is transferred into characters' minds, perfecting the transmuted object by dyeing it with various colours of figurative language. Bloom, a wanderer about Dublin on a normal warm summer day, encounters a network of Dublin acquaintances while Dublin's cityscape serves as a vibrant cobweb of stimulants, triggering multiple intersecting planes of thoughts in his mind.

The rambling monologue and the hard-to-follow fragmentation result in continuous disruption of the spatial narrative. Being introduced to the familiar space of Dublin at first, the reader is then transferred to totally strange spaces in Bloom's mind. The external reality is mixed with the internal one, while Dublin is being transmuted into an intermingled entity, simultaneously real and unreal. An example could serve to illustrate how this technique works:

> He crossed at Nassau street corner and stood before the window of Yeates and Son, pricing the fieldglasses. Or will I drop into old Harris's and have a chat with young Sinclair? Wellmannered fellow. Probably at his lunch. Must get those old glasses of mine set right. Goerz lenses six guineas. Germans making their way everywhere. Sell on easy terms to capture trade. Undercutting. Might chance on a pair in the railway lost property office. Astonishing the things people leave behind them in trains and cloakrooms.[33]

The real place name, Yeates and Son, informs the reader of Bloom's locality, and immediately he is transferred to the jewellery store in the character's mind to meet Sinclair and imagine him having lunch, then to a series of Bloom's plans and reflections such as the intention to buy a new glass, or the quality of German goods.

Innovative literary techniques arise out of experimentation due to the need to catch up with the unpredictable reality, and Joyce's interior monologue proves true his statement that changes in life lead to changes in style. The changes in life that Joyce is aware of spring from the modern city with its numerous facets and the way it affects the human mind. In his influential analysis of the effects of the metropolis on the mental life, Georg Simmel proposes that 'The psychological basis of the metropolitan type of individuality consists in the intensification of nervous stimulation which results from the swift and uninterrupted change of outer and inner stimuli'.[34] The new 'rhythm of life' with its 'onrushing impressions' affects the mind of the citizen tremendously.[35] With each passing by streets, stores, squares, Dubliners by Bloom, the modern city of Dublin, 'with the tempo and multiplicity of economic, occupational and social life', a contrast to the rural life, strikes his mind with tremendous force and creates a flood of interior thoughts.[36] While walking, Bloom receives thousands of stimuli on his sensory system and interior monologue is invoked to report these mental processions. The technique reveals how the human brain records quick and immediately departing impressions, creating in the character's mind a running flow of thoughts.

Franco Moretti in his reading of *Ulysses* as a modern epic also discusses Simmel's theory to propose his explanation about how interior monologue is a technique invented by Joyce, a city writer. This style consists of 'a weaker grammar than that of consciousness; an edgy, discontinuous syntax: a cubism of language' and 'simple, fragmented sentences', 'paratactical paragraphs'.[37] Beyond any doubt, Bloom's inner mind is represented by a collection of sentences of broken syntaxes: for example, 'God. Save. Our.',[38] 'Torry and Alexander last year',[39] 'Rough weather outside',[40] 'If he…? O! Eh? No … No'.[41] These subjectless sentences are jotted down as if coming straight from Bloom's mind. Furthermore, interior monologue works side by side with third-person narrator and the montage technique, deftly employed by Joyce to depict Bloom's movements and to blend them with his mind, underlining their coexistence. The successive ideas that run through Bloom's mind are situated one after another in interrelated broken sentences. This method reveals how one subject is seen from various views and highlights the multiplicity of life in general. This diversity of both reality and mind is carried out by the repeated changes of focalisation which progress from omniscient narration into interior monologue. In a very short time, the reader is shown Bloom watching the Yeats and Son store, then suddenly considering a new pair of glass, then turning his thoughts to how German goods are traded everywhere. The city of Dublin is no longer an objective thing set up in a linear sequence of realist descriptions, but is now dissected into numerous images simultaneously present. No longer borne of static representation, Dublin is depicted in moving montages, calling attention to its textual heterogeneity.

Despite 'the additive nature of language', Joyce still manages to create the 'the effect of multidimensional reality' by continuously jamming in a series of visual and aural descriptions.[42] Moreover, textual multiplicity is produced by the rise and fall of Bloom's emotions that informs the style. As a result, Joyce recreates a whole new Dublin corresponding to Bloom's currently reigning condition: a food-related Dublin in 'Lestrygonians' and so on. Specifically, Bloom's states of mind are attuned to his states of body, explaining why Joyce himself calls his book 'the epic of the human body': 'In "Lestrygonians" the stomach dominates and the rhythm of the episode is that of the peristaltic movement'.[43] Similar to how Bloom walks toward his lunch thinking of his wife's legs as 'out of plumb', subtly referring to the plum fruit, Bloom's hunger furnishes the prose with suitable language. At this 'worst hour of the day', food allusions are everywhere and the physical world of Dublin is transformed into an edible and hungry city as in Charles Peake's sharp discernment: the 'gaunt' and 'famished' gulls flying between the quay walls; the houses in Western Dublin become 'mushroom'; the stone of the National Museum has 'cream curves'; poplin in Brown Thomas' store is transformed into 'a flood of bloodhued' and 'lustrous blood'; and a clock in Davy Byrne's throws Bloom a 'bilious' look.[44]

In *Ulysses*, the process of Dublin being transferred into the psyche of the character happens more smoothly, due to the gracious blending of omniscient narration and interior monologue. The fluidity by which the narration moves between fictional and historical geography, internal and external reality, is strengthened in Bloom's episodes through the interior monologue with its incomplete sentences, noun phrases, and broken syntaxes to render non-logical free associations mimicking the way impressions of the world fall upon and are randomly caught in the character's mind. The city touches on and explores Bloom's private buried aspects of mind, activating, awaking deep dormant feelings, submerged thoughts, and concealed memories. Local places in Dublin contain personal memories of its citizens and each meeting prompts a series of reminiscences. The city becomes the site of semantic chains of signification, in which place names serve in turn as signifiers and signifieds. Specifically, from a theoretical stance, Barthes discusses how signifieds and signifiers work in the site of a city:

> the signifieds [...] always become the signifiers of something else: the signifieds pass, the signifiers remain. [...] the signifieds are always signifiers for other; and reciprocally. In reality, in any cultural or even psychological complex, we find ourselves confronted with infinite chains of metaphors whose signified is always recessive or itself becoming a signifier.[45]

Following this cue, the reader witnesses this chain of signification in Dublin in every passage:

> He crossed Westmoreland street when apostrophe S had plodded by. Rover cycleshop. Those races are on today. How long ago is that? Year Phil Gilligan died. We were in Lombard street west. [...] Happy. Happier then. Snug little room that was with the red wallpaper, Dockrell's, one and ninepence a dozen. Milly's tubbing night.[46]

Westmoreland starts as a signifier, a signification in the free association system. Crossing it, Bloom thinks of Rover cycleshop, a signified. In its turn, the Rover cycleshop becomes a signifier, referring to the races, and in this way, places refer to places refer to people refer to events refer to times in the past refer to places in the past. Signifiers become signifieds and signifieds become signifiers. Lombard street west is the signified of the signifier that is 'Year Phil Gilligan died', in its turn it signifies multiple memories in Bloom's past, the most important of which is the happier time of his life. The golden time of the Blooms before the death of their son is revived and re-performed onto present vividly thanks to the mere act of crossing a street. Thereby, within interior monologue itself, there is a chain of signifiers and signifieds. The device, with its initial and primary function being to reflect reality, develops beyond to create a semantic field. The chain of signifieds, facilitated by the language of interior monologue, ceases to correspond to reality: it cuts itself off from the source in reality and becomes an assortment of floating signifieds and signifiers. Dublin is transformed into a site of signification, 'in which the topographical features of the physical city combine with the mental landscape created by its inhabitants to create a complex and paradoxical world of signification'.[47]

More importantly, Bloom's 'world of signification' is where the past and the present are conflated. Due to the technique of simultaneously blending omniscient narration with interior monologue, an encounter with the present mnemonically induces a remembrance of things past. The body of the character receives a stimulus and the mind works in a mysterious way to make two distinct worlds subsume each other. The body of the character exists in two worlds: it makes the reader (and the character himself) experience an illusion, a perception disorder; for a split moment, he is in the past and yet not in the past, he is in the present and yet not in the present. The stimulus might be a sound, a sight, or a smell. In the above passage, while the Bloom of the present is crossing Westmoreland Street, the Bloom of the past occupies his stream of consciousness where his happier days are performed and come full circle in his mind. Momentarily, the happier days with Milly as a kid, the rabbit pie, Molly's dress, and the present world of Bloom passing a street absorb each other and are conflated into one. The signifiers become signifieds, and the chain goes on forever until the reader cannot distinguish which triggers which any longer. The world of Lombard street west is evoked once more when Bloom is passing Adam court: 'I was happier then. Or was that I? Or am I now I? Twentyeight

I was. She twentythree when we left Lombard street west something changed'.[48] In addition, Bloom asks himself: 'Can't bring back time. Like holding water in your hand. Would you go back to then?'[49] He cannot bring back time, but his involuntary memory receiving several cues in his ramble in the city can bring back the vivacity of the past in full force. There is a spatial dimension of experience, and *Ulysses* itself moves the reader in and out of various mental 'planes' from moment to moment. The Dublin montage defies any pure and single dimension of space and advocates for a multiplicity of places, objective as well as figurative.

The text, at one level, signifies 'this is reality' with proper names, tempting the reader to a second level, a metaphor of reading. As a result, the reader follows the text's recording of Bloom's movements all around the city, attempting to find significations in them and understand the world and yet experiencing problems because his wanderings are a journey without end. The text is invaded by the cultural code and bears too many signifiers; thus, it exhausts semantic exploration: social and political symbolic interpretations are always intersecting with Bloom's personal memories.[50] Place names in *Ulysses* are abundant with their illusory effect of reality: they produce the objectivity of Dublin, leading the reader in following the character's steps to not only encounter the outer reality, the city of Dublin, but also the figurative mind. The act of listing proper names becomes the act of deferring meaning and the production of meaning is delayed. At another level, the city shrinks into each place name listed by the narrator, but the very next moment it spans out tremendously impelled by memories. Stream of consciousness challenges the realness of the discourse, preventing the narrative of realness from taking centre stage: the moment a real place name gets mentioned, the reader is promptly taken to the space of internal thoughts. The world of Dublin slips away to somewhere else, to the space of heterogeneity and multiplicity. The places in Bloom's memories are of changeable impressions, of Howth being under the sun and a seedcake sharing in Molly's mouth, of Davy Byrne's being stuck in two glass planes and a gulp of wine awakening, fleeting moments and places. They are not distinguishable and sharply separated moments and places as we might think. Howth and Davy Byrne's, the happy courting day and now, all dissipate, are no longer specific, material, and separate places and time as we often conceive of them. Rather, in a moment, they absorb each other, crystalising and becoming ephemeral.

Dublin's landscape is not the space and place of the map but of Bloom's life, veiled fragments of transitory memories and feelings, involuntarily reawakened and contorted. Thus, there is an ongoing struggle between these two levels, and *Ulysses* urges the reader to reflect upon the relationship between reality and art, the real and the fictive Dublin: Joyce's Dublin is similar to Derrida's *différance*, that is 'neither this nor that; but rather this and that [...] without being reducible to a dialectical logical either' (Derrida, *Deconstruction and the Other* 161). It rejects being a realistic and symbolic city and yet never quite leaves the realm of hyperreality of Dublin; it forms itself in Bloom's mind and projects itself onto the outer world of Dublin: the decentralised textual Dublin, whose nature is '[l]anguage and a finite language—[thus it] excludes totalisation',[51] with its multiplicity reveals how Joyce's writing defies imposed traditions of literary reading and criticism, as well as meaning and coherence.

Notes

1. James Joyce, *Letters of James Joyce*, vol. 2, ed. Richard Ellmann (New York: Viking, 1966), 134.
2. Joyce, *Letters of James Joyce*, vol. 1, ed. Stuart Gilbert (New York: Viking, 1957), 175.
3. Stewart Bowers et al., 'Oliver Twist – Mapping Literary London,' accessed 22 January 2016, http://mappingliterarylondon.weebly.com/oliver-twist.html.
4. Jonathan D Culler, 'Barthes, Theorist,' *The Yale Journal of Criticism*, vol. 14, no. 2 (Fall 2001): 440.
5. They are proairetic code, hermeneutic code, connotative code, and symbolic code.
6. Roland Barthes, *S/Z*, trans. Richard Miller (New York: Noonday P, 1974), 20.
7. Philippe Hamon, 'On the Major Features of Realist Discourse,' in *Realism*, ed. Lilian R. Furst (New York: Longman, 1992), 168.
8. Hamon, 'On the Major Features of Realist Discourse,' 168.
9. Benjamin Hrushovski, 'Fictionality and Fields of Reference: Remarks on a Theoretical Framework,' *Poetics Today* 5.2 (1984): 244.
10. Martin Heidegger, *What is Called Thinking*, trans. J. Glenn Gray and F. Wieck (New York: Harper & Row, 1968), 119.
11. Terry Eagleton, *The English Novel: An Introduction* (Malden, MA: Blackwell Publishing, 2005), 10.
12. Frank Budgen, *James Joyce and the Making of 'Ulysses,' and Other Writings* (London: Oxford University Press, 1972), 70.
13. Elaine Scarry, *Dreaming by the Book* (New York: Farrar, Straus and Giroux, 1999), 9.

14. Quyen Nguyen, '"Dublin what place was it': Making sense of the textual city in *Ulysses*', *MoveableType*, Vol. 9, 'Metropolis' (2017): 4. It could be said that Frank Budgen in *James Joyce and the Making of Ulysses* sowed the seeds of a geographically focused approach in Joyce's criticism. Following his lead, Dublin in *Ulysses* has been consistently compared and conflated with the real city. See earlier works on this topic: Richard Kain, *Fabulous Voyager: A Study of James Joyce's "Ulysses"* (Chicago, IL: University of Chicago Press, 1947), Robert Martin Adams, *Surface and Symbol: The Consistency of James Joyce's "Ulysses"* (New York: Oxford University Press, 1962), Clive Hart and Leo Knuth, *James Joyce's Dublin – A Topographical Guide to the Dublin of "Ulysses"* (Colchester: A Wake Newslitter Press, 1975). Many recent articles discuss and compare Joyce's Dublin to Dublin in 1904. See Enda Duffy, 'Setting: Dublin 1904/1922' in *The Cambridge Companion to Ulysses*, ed. Sean Latham (Cambridge: Cambridge University Press, 2014), 81–94; Terence Killeen, 'Lee Miller: Photographing Joycean Dublin (1946)' in *Voices on Joyce*, ed. Anne Fogarty and Fran O'Rourke (Dublin: University College Dublin Press, 2015), 133–136.
15. David Spurr, *Architecture and Modern Literature* (Ann Arbor: University of Michigan Press, 2012), 188.
16. Andrew Thacker. *Moving through Modernity: Space and Geography in Modernism* (Manchester: Manchester University Press, 2003), 122.
17. Culler explains that 'Recuperation is a process of making details into ***signifiants*** [sic] and naming their ***signifies*** [sic]. The drive towards meaning on the reader's part is extremely powerful, and, as we shall see, the ubiquity of the cultural models and symbolic codes which guide it makes it a difficult process to disrupt.' Culler, *Flaubert: the Uses of Uncertainty*, 62.
18. Culler, *Flaubert: the Uses of Uncertainty*, 72.
19. Joyce, *Ulysses*, 160.
20. Joyce, *Ulysses*, 161.
21. Joyce, *Ulysses*, 161.
22. Joyce, *Ulysses*, 167.
23. Joyce, *Ulysses*, 171.
24. Erwin Ray Steinberg, *The Stream of Consciousness and Beyond in 'Ulysses'* (Pittsburgh: University of Pittsburgh Press, 1973), 95.
25. Scarry, *Dreaming by the Book*, 31–36.
26. Joyce, *Ulysses*, 151.
27. Joyce, *Ulysses*, 167.
28. Joyce, *Ulysses*, 168.
29. Karen Lawrence, *The Odyssey of Style in 'Ulysses'* (Princeton, NJ: Princeton University Press, 1981), 45.
30. Monica Fludernik also spots this tendency of aloofness in Bloom's episodes. She notes an 'independent narrative,' with a distancing effect, is a

result of the 'tendency towards pedantry and the furnishing of circumstantial detail,' hence 'establishing a distinct narrative voice emancipated from Bloom's point of view'. Monica Fludernik 'Narrative and Its Development in *Ulysses*', *Journal of Narrative Technique 16*, i (Winter 1986), 17.

31. Barthes, 'Semiology and Urbanism' in *The Semiotic Challenge*, trans. Richard Howard (Oxford: Basil Blackwell, 1988), 195.
32. Barthes, 'Semiology and Urbanism', 199.
33. Joyce, *Ulysses*, 166.
34. George Simmel, 'The Metropolis and Mental Life' in *The Sociology of Georg Simmel* (New York: Free Press, 1976), 409.
35. Simmel, 'The Metropolis and Mental Life', 409.
36. Simmel, 'The Metropolis and Mental Life', 410.
37. Franco Moretti, *Modern Epic: The World-System from Goethe to García Márquez* (London; New York: Verso, 1996), 134–135.
38. Joyce, *Ulysses*, 151.
39. Joyce, *Ulysses*, 151.
40. Joyce, *Ulysses*, 152.
41. Joyce, *Ulysses*, 153.
42. Steinberg, *The Stream of Consciousness and Beyond in 'Ulysses'*, 52.
43. Budgen, *Making of 'Ulysses,'* 21.
44. Charles Peake, *James Joyce, the Citizen and the Artist* (Stanford: Stanford University Press, 1977), 200.
45. Barthes, 'Semiology and Urbanism', 199.
46. Joyce, *Ulysses*, 155.
47. Desmond Harding, *Writing the City: Urban Visions and Literary Modernism* (London: Routledge, 2003), 34–35.
48. Joyce, *Ulysses*, 168.
49. Joyce, *Ulysses*, 168.
50. Roland Barthes, 'Semiology and Urbanism,' 198.
51. Jacques Derrida, 'Structure, Sign, and Play in the Discourse of the Human Sciences' in *Writing and Difference*, trans. Alan Bass (London: Routledge & Kegan Paul, 1981), 365.

References

Adams, Robert Martin. *Surface and Symbol: The Consistency of James Joyce's "Ulysses."* New York: Oxford University Press, 1962.

Barthes, Roland. *S/Z*. Translated by Richard Miller. New York: Noonday Press, 1974.

———. 'Semiology and Urbanism.' In *The Semiotic Challenge*, trans. Richard Howard, 191–201. Oxford: Basil Blackwell, 1988.

Bowers, Stewart, et al. 'Oliver Twist—Mapping Literary London.' http://mappingliterarylondon.weebly.com/oliver-twist.html. Accessed January 22, 2016.

Budgen, Frank. *James Joyce and the Making of 'Ulysses,' and Other Writings.* London: Oxford University Press, 1972.

Culler, Jonathan D. *Flaubert: The Uses of Uncertainty.* Ithaca, NY: Cornell University Press, 1974.

———. 'Barthes, Theorist.' *The Yale Journal of Criticism* 14, no. 2 (Fall, 2001): 439–446.

Derrida, Jacques. 'Signature Event Context.' In *Glyph*, vol. I, 172–197. Baltimore: Johns Hopkins University Press, 1977.

———. 'Structure, Sign, and Play in the Discourse of the Human Sciences.' In *Writing and Difference*, trans. Alan Bass, 351–370. London: Routledge & Kegan Paul, 1981.

———. 'Deconstruction and the Other.' In *States of Mind: Dialogues with Contemporary Continental Thinkers*, ed. Richard Kearney, 156–176. Manchester: Manchester University Press, 1995.

Duffy, Enda. 'Setting: Dublin 1904/1922.' In *The Cambridge Companion to Ulysses*, ed. Sean Latham, 81–94. Cambridge: Cambridge University Press, 2014.

Eagleton, Terry. *The English Novel: An Introduction.* Malden, MA: Blackwell Publishing, 2005.

Fludernik, Monica. 'Narrative and Its Development in Ulysses.' *Journal of Narrative Technique* 16, no. i (Winter, 1986): 15–40.

Hamon, Philippe. 'On the Major Features of Realist Discourse.' In *Realism*, ed. Lilian R. Furst, 166–185. New York: Longman, 1992.

Harding, Desmond. *Writing the City: Urban Visions and Literary Modernism.* London: Routledge, 2003.

Hart, Clive, and Leo Knuth. *A Topographical Guide to James Joyce's "Ulysses."* 2 vols. Colchester: A Wake Newslitter Press, 1975.

Heidegger, Martin. *What Is Called Thinking.* Translated by J. Glenn Gray and F. Wieck. New York: Harper & Row, 1968.

Hrushovski, Benjamin. 'Fictionality and Fields of Reference: Remarks on a Theoretical Framework.' *Poetics Today* 5, no. 2 (1984): 227–251.

Joyce, James. *Ulysses.* New York: Random House, 1934; reset and corrected 1961.

———. *Letters of James Joyce.* Vol. 1. Edited by Stuart Gilbert; Vols. 2–3. Edited by Richard Ellmann. New York: Viking, 1957, 1966.

Kain, Richard M. *Fabulous Voyager; James Joyce's "Ulysses."* Chicago, IL: University of Chicago Press, 1947.

Killeen, Terence. 'Lee Miller: Photographing Joycean Dublin (1946).' In *Voices on Joyce*, ed. Anne Fogarty and Fran O'Rourke, 133–136. Dublin: University College Dublin Press, 2015.

Lawrence, Karen. *The Odyssey of Style in 'Ulysses.'* Princeton, NJ: Princeton University Press, 1981.

Moretti, Franco. *Modern Epic: The World-System from Goethe to García Márquez.* London; New York: Verso, 1996.

Nguyen, Quyen. '"Dublin What Place Was It": Making Sense of the Textual City in *Ulysses*'. *MoveableType*, Vol. 9, 'Metropolis' (2017): 4–19.

Peake, Charles. *James Joyce, the Citizen and the Artist.* Stanford: Stanford University Press, 1977.

Scarry, Elaine. *Dreaming by the Book.* 1st ed. New York: Farrar, Straus and Giroux, 1999.

Simmel, George. 'The Metropolis and Mental Life.' In *The Sociology of Georg Simmel*, 409–424. New York: Free Press, 1976.

Spurr, David. *Architecture and Modern Literature.* Ann Arbor: University of Michigan Press, 2012.

Steinberg, Erwin Ray. *The Stream of Consciousness and Beyond in "Ulysses."* Pittsburgh: University of Pittsburgh Press, 1973.

Thacker, Andrew. *Moving through Modernity: Space and Geography in Modernism.* Manchester: Manchester University Press, 2003.

CHAPTER 7

Urban Degeneracy and the Free State in Flann O'Brien's *At Swim-Two-Birds*

Laura Lovejoy

> *Purporting to be an immoral character, I accompanied him on a walk through the environs of Irishtown, Sandymount, and Sydney Parade, returning by Haddington Road and the banks of the canal.*[1]

As the diverse chapters in this volume illustrate, the city in Irish fiction is often an ambivalent space with the power to both elevate and debase its inhabitants. Indeed, portrayals of urban squalor, seediness, and corruption, while prevalent in Irish fiction, have been balanced with depictions of the city as a space of positive growth and transformation.[2] Modernist urban fictions in particular have captured the city's dual forces of progress and regression.[3] The Dublin of James Joyce's *Ulysses* (1922), for example, perhaps the most iconic work of Irish urban fiction, is at once a secretive space of vice, corruption, and anonymous desire and a forum of spectacle which showcases futuristic aspects of modernity, resisting what Joyce saw as nationalist provincialism. In its complexity and multifacetedness, Joyce's portrayal of Dublin, although formally radical, approaches a kind of realism, exemplifying the modernist tendency to use avant garde representational techniques to capture the sociocultural realities of the early twentieth-century city.[4] Joyce's high modernist 'aesthetic of comprehensiveness and

L. Lovejoy (✉)
University College Cork, Cork, Ireland
e-mail: laura.lovejoy@ucd.ie

M. Beville, D. Flynn (eds.), *Irish Urban Fictions*, Literary Urban Studies, https://doi.org/10.1007/978-3-319-98322-6_7

minute representation' in relation to the urban was so meticulous that he claimed Dublin could be reconstructed from his book were it to disappear.[5] *Ulysses* is a useful starting point for considering the function of the urban in Irish modernist fiction, not least because Flann O'Brien's and Joyce's fictive renderings of the urban continue to invite comparisons. More importantly, however, *Ulysses*' representative authenticity in relation to the urban serves as a contrasting precursor to the more superficial and burlesque depiction of post-independence Dublin with which this chapter is concerned.

Sharing with *Ulysses* its setting of Dublin, O'Brien's *At Swim-Two-Birds* (1939) is also an urban Free State novel. Composed during the Free State and published two years after the implementation of the constitution and The Free State's transition to *Eire* in 1937, the text responds to the specific cultural and political milieu of Dublin in the 1930s, as the work of scholars such as Declan Kiberd and Joseph Brooker has illustrated.[6] Yet, despite the novel's clear correspondence to an identifiable sociohistorical milieu, it does not offer what can be termed a realist portrayal of Dublin in the 1930s. Indeed, the text's late modernist distortion of novelistic conventions and its surrealist intra-narratives signal its overall departure from realist modes. The absence of Ulyssean realism in *At Swim-Two-Birds* can be attributed in part to a shift in modernist aim and aesthetics. Unlike his predecessors, O'Brien's aim is not to use fiction to capture the essence of the postcolonial city, but rather to use the city and its cultural associations to satirise contemporary anxieties about Irish moral and cultural degeneration. If the Dublin of *Ulysses* is multifaceted, then O'Brien's Dublin has one primary facet—degeneracy.

At Swim-Two-Birds exaggerates pre-existing cultural associations between the urban and the immoral to comic effect. In *At Swim-Two-Birds*, the city's main thoroughfares and side-streets, and its public and hidden places, offer seemingly endless opportunities for overriding the morality of those who were invested in a Catholicised vision of Ireland's post-independence purity. The fictionalised city's moral topography provides a subversive counterpoint to the visions of national purity disseminated by architects of the post-independence cultural landscape, from which, O'Brien's late modernist imagination evolved. Traversing the city, the narrator is seduced by urban cultures of drinking, gambling, and loitering, and the cosmopolitan appeal of the picture-house. Representing

Dublin as a forum of modern iniquity, O'Brien constructs a caricature of a city and, in doing so, plays with the city's status as a spatial focal point of Free State anxieties surrounding urban-centred immorality. Probing the city's implication in discourses of national decline in the Irish Free State, this chapter resituates O'Brien's satiric representation of the city as a morally corrupting space in *At Swim-Two-Birds* in relation to contemporary discourses of Irish cultural and moral degeneration.

Moral and Immoral Environs in *At Swim-Two-Birds*

Space is linked with morality in *At Swim-Two-Birds*. This is obvious at the novel's opening when the narrator is chastised by his uncle for maintaining an air of secrecy around his activities by secluding himself in his bedroom: 'I know the studying you do in your bedroom, said my uncle. Damn the studying you do in your bedroom'.[7] O'Brien later emphasises this association between the bedroom and moral impropriety when the narrator's friend Brinsley remarks that the narrator's 'curious bedroom smell'[8] is reminiscent of 'a room early in the morning ... where there had been a hooley the night before, with cigars and whisky and crackers and women's scent'.[9] The more morally sanitised, wholesome zones of the homestead, such as the kitchen and sitting room, are, by contrast, associated with purity, and O'Brien uses these spaces to parody the fervent preoccupations of the Gaelic League, an organisation founded in the late nineteenth century which encompassed morality in its objective of promoting the Irish language and Irish culture. As the narrator's uncle and his Gaelic League friends hold a meeting to plan a *ceilidh* in the kitchen, they are divided over issues such as the Irishness or un-Irishness of certain dances.

If O'Brien draws on Free State associations between domesticity and purity in order to lampoon fetishistic nationalism, so, too, does he deploy the city and its connotations of degeneracy to ridicule the alarmist sensibilities of those who subscribed to narratives of national moral and cultural decay. The urban space of Dublin, even more so than the private sanctum of the bedroom, is coded as immoral in the novel. Indeed, the Dublin of *At Swim-Two-Birds* is one of the most superficially decadent cities in literary modernism. Descriptions of the city are primarily embedded in the first narrative layer of the novel—the introductory framework which follows the literary enterprises of an under-achieving undergraduate (the narrator) during a year of study at University College Dublin in the early 1930s. As the student is increasingly distracted by the thrills afforded by metropolitan life, he is lured further from the vision of virtuous manhood valorised by his uncle, a devotee of the Irish Christian Brothers and a caricature of Catholic male authority in early twentieth-century Ireland.

The city centre's 'glistening'[10] streets offer the narrator solace from the prying eyes of his uncle, who is prone to imposing moralistic platitudes about 'Godgiven health' and the ideal qualities of 'boys'.[11] The uncle's role as a token of Catholic surveillance is made explicit at the novel's opening, when he insists on supervising the narrator's activities, asking 'Why don't you study in the dining room here where the ink is and where there is a good book-case for your books? Boys but you make a great secret of your studies'.[12] As Gregory Dobbins has noted, the uncle is 'the most obvious figure of authority in the novel'.[13] Yet, he is more than a paternalistic emblem positioned for rebellious youth to rage against; in his scrutiny, near omnipresence, and vehement Catholicism, he comes to stand for the authority of the Catholic Church and the Irish Free State, forces which are mocked at every turn in the text.

Even more so than the secretive space of his bedroom, the privacy of which is shattered by the uncle's routine impositions, the city obscures the narrator's movements, its topography, and personae, providing cover from the penetrating gaze of Free State Catholicism. For O'Brien, the city becomes a space which facilitates the evasion of authority, a notable departure from the Dublin of *Ulysses*, where citizens are confronted with authority in the form of colonial architecture and must navigate the material culture of the state. In the city of *At Swim-Two-Birds*, the Church's grip on the narrator is loosened progressively, as he increasingly embodies the kind of boy his uncle condemns at the start of the novel: a boy 'who gives himself up to the sin of sloth … one of the worst of the deadly sins'.[14]

The uncle's variety of supervisory morality reflects the fact that surveillance of citizens was a core feature of the Irish Free State's regulation of sexuality and morality following independence. Research by Sandra L. McAvoy and Louise Ryan has shown the extent to which religious, journalistic, and legislative forces were collusive in perpetuating an atmosphere of moral scrutiny in the service of countering sexual and other forms of 'immorality' in the 1920s and 1930s.[15] Moral surveillance characterised to an intensified degree the sociocultural milieu O'Brien captures and satirises in *At Swim-Two-Birds*, as did pervasive beliefs about the decline of Irish society, and O'Brien's novel is keenly attuned to the fact that such anxieties about moral and cultural decline were often focused on the urban.

DEGENERATE DUBLIN

While nationalist tropes of Irish cultural supremacy had long presented the rural as a zone of unsurpassed purity and tradition, and the urban, by contrast, as a danger zone, in the Free State, Dublin was increasingly framed as the moral Achilles' heel of Ireland. Owing to its geographical and supposed cultural proximity to England and its comparatively cosmopolitan status, Dublin appeared to threaten nation-building efforts, which were rooted in the pursuit of national self-sufficiency and moral purity, by encouraging over-indulgence in leisure and vice. Indeed, the increasingly sordid view of Dublin emerging in degeneration discourses in the 1920s and 1930s appeared to contradict the Christian Brothers' 1924 declaration that 'purity is the national virtue of Ireland'.[16] Furthermore, the immorality apparently concentrated in Dublin during the Free State years was in danger of corrupting not just urbanites, but spreading throughout the country and defiling the historically more pure and wholesome rural stock.[17]

Characterisations of the city as a sinister space reflected growing political concerns about Ireland's future cultural and economic stability and linked the degeneration of Irish life not just to the apparent increase in crime in Irish cities but to the overabundance of leisure and sin. The writer and historian Patrick Sarsfield O'Hegarty was one of a number of nationalists who linked the 'craze for idleness and amusement' emerging in Irish cities with societal decline, writing in *The Victory of Sinn Féin* (1924) that Irish society must 'get back to simplicity and strenuousness' and resist the dazzling allure of the city 'or we shall certainly go out'.[18] O'Hegarty's comments on urban leisure reflect the fact that, from the early 1920s onward, concerns about Ireland's supposed moral and cultural decline took on a notably spatial aspect. The Catholic Church and associated groups attempted to counter what they viewed as a growing crisis in relation to immorality, which was apparently fostered by expressions of modernity flourishing in urban areas, through the circulation of instructive and regulatory discourses.

Indeed, many of the most vehement attempts to counter the perceived degeneration of Irish culture and morality came from nationalist and Catholic groups, one of the most active and prolific of which was the Irish Christian Brothers. A group whose brand of self-regulatory didacticism was deeply familiar to O'Brien, their publication *Our Boys* targeted the moral debasement of youth through the promotion of Catholic virtues of 'conscience … self-training, contentment, purity, and duty'.[19] While the Christian Brothers instilled in 'their boys' the importance of education, nutrition, athleticism, and devotion to God, the pages of their magazine devoted equal, if not more, space to condemning deplorable behaviours, betraying, as Daire Keogh has noted, their fears that Catholic values were being rejected in favour of the thrill of urban-centred modern entertainments.[20]

A 1922 issue declared that one of the objectives of *Our Boys* was 'To preserve them (the children of the Gael Irish) from those withering influences that lead to Moral and National Decay'.[21] Such decaying influences were, it is notable, often encountered in 'the streets'. A letter in the same issue, for example, extolling the virtues of manual labour, urged boys to keep away from 'the streets' where 'withering influences' could take hold, its 11-year-old author noting that hobbies such as fret-work 'will help to keep them [boys] away from the street' and adding that the Christian Brothers' magazine had helped him to give up smoking.[22]

The concerns expressed by the Christian Brothers about urban degeneracy were part of a wider cultural narrative of national decline which was played out across a range of print media including national and regional press in the Free State. Panicked reporting on the vices supposedly concealed in Dublin's inner-city throughout the 1920s and 1930s magnified the public focus on Ireland's cultural decay. Disparate articles reveal the extent to which Dublin, more than any other city or town in the Free State, was characterised in journalistic discourse as a location of immorality. As Ciara Breathnach notes, urban life was 'demonised and feared' in the Free State.[23] Dubliners themselves believed that immorality was festering in their city, as letters to the editor of the Evening Herald in the 1920s show. One contributor complained that there were 'at present in the city of Dublin shops displaying immoral figures, pictures, etc.'.[24] Another correspondent was alarmed to find that 'Most of the productions shown in Dublin are absolutely rotten with immorality'.[25] A 1931 article in the *Irish Independent* noted that local government reports had apparently 'supplied evidence of sexual sin and a debased moral standard on an alarming scale' and that 'An ever darker side of the low moral standard now prevailing in Dublin city' was 'revealed by the proceedings of the courts of justice'.[26]

Those concerned with Ireland's rapidly eroding morality turned their attention to pre-existing social problems and vices. Despite its longstanding existence in the capital, gambling became framed as a new social evil which was increasing at an alarming rate and emanating from Dublin.[27] It is difficult to ascertain the extent to which claims that gambling was increasing were true. Yet, what is significant is the degree to which the catastrophising of urban-centred forms of leisure, such as gambling, was a feature of nationalist rhetoric in the Free State and shaped the milieu O'Brien responds to in *At Swim-Two-Birds.* Gambling was framed specifically as an urban problem. An *Our Boys* article from 1927, for example, warned that 'Gambling is admittedly one of the greatest, if not the greatest, of our national evils'. The article's author, 'an old boy of the Christian Brothers', declared that first Ireland's 'large cities' and then 'every town and village in the country' were 'suffering from the blighting influence of the gambling passion' and that efforts were desperately needed to 'eliminate the betting curse from our midst'.[28] Gambling contravened the Christian Brothers' principles of resolution and diligence as it promised boys 'something for nothing' and encouraged urban leisure rather than the simplicity and strenuousness P.S. O'Hegarty urged.[29] The impressionable narrator of *At Swim-Two-Birds*, as I will discuss, has evidently been infected by Dublin's gambling craze.

Articles in the national press, too, addressed the idea that Dublin was a haven for betting establishments. In the late 1920s, a proliferation of articles referred to Dublin's 'gambling' and 'betting craze'.[30] Citizens from outside of Dublin regarded it as a hive of nefarious gambling activity. An article in the Ulster Herald from 1928 noted that 'In Dublin the activities of betting offices had grown to such an extent that the police authorities were pressing for fresh legislation ... Horse-racing, formerly a fine and noble sport, was now dominated by gambling, which was also demoralising football, and many other games'.[31] Sensationalist articles about the dangers of Dublin-based gambling continued into the 1930s. A 1930 article in the Irish Independent, for example, described Dublin's 'Gambling Hells' and their culpability in the moral debasement of young men, reporting a Dublin solicitor's claim that 'More crimes have been committed since these betting houses opened—gambling hells I call them'. Others echoed tropes of urban chaos and disorder, noting that such establishments were 'cluttered up in the purlieus of the city'.[32] Protestant leaders also decried the corrupting influence of betting, pointing to Dublin as a bad moral influence on the wider community. In 1931, Archdeacon MacNeice referred to the 'gambling craze' and claimed that the 'Dublin craze for sweepstakes is a menace to the well-being of the whole community',[33] and in 1933, Reverend W.H. Forde noted that 'In Dublin there was a truly demoralising spectacle which cast a gloom over the land'.[34]

Another social evil framed as a specifically urban problem was drink. Intemperance had long been subject to organised intervention by social purity groups in Ireland, as work by Katherine Mullin has shown.[35] Yet, as with gambling, discourses constructing drinking as a national problem were intensified in the Free State, and intemperance was cited as evidence of Ireland's moral decline. The problem of national drunkenness was countered with abstinence campaigns. For success, the Christian Brothers insisted upon 'rigid abstention from alcohol and tobacco' and stated that 'self-control in pleasures—e.g., smoking, drink, theatres, clubs, etc., indicates great respect for yourself, your health, those depending on you for happiness, etc.'.[36] Dublin was, unsurprisingly, the place most in need of temperance efforts according to the Brothers, who claimed in 1927 that 'if the familiar saying, that where Dublin leads the country follows, be true, then we may all live to see the ugly fetter of intemperance struck for good from the soul of Ireland'.[37] The Christian Brothers included temperance manifestos in their educational materials as well as their magazines. Indeed, an excerpt from an essay entitled 'Alcohol and Health' from the Christian Brothers' Higher Literary Reader (1925) is reproduced in its entirety in *At Swim-Two-Birds.* Before the narrator describes his first encounter with and subsequent immediate addiction to alcohol, he includes an extract from a 'schoolbook which [he] read at the age of twelve' which describes alcohol as 'an irritant and a narcotic poison' which 'excites the brain, quickens the action of the heart, produces intoxication and leads to degeneration of the tissues'.[38,39]

Dublin was also contributing to 'base denationalization' through its association with immoral literature.[40] In 1922, when O'Brien was 11, the editor of *Our Boys* launched an appeal to its readers to resist the foreign and degenerate literature being circulated on the streets of Dublin, writing that 'There are at present ship loads upon ship loads of immoral, filthy, and unhealthy Cross-Channel Literature pouring into this city … Voluntary and earnest workers are wanted to start crusades in all parts of Dublin'.[41]

As a letter to the editor of *Our Boys* in 1927 illustrates, Dublin's streets were seen to expose Ireland's youth to the 'dregs of fiendish literature … papers, periodicals' and 'foul magazines'.[42] Irish nationality, as well as morality, the author noted, had 'been blighted to an abnormal extent by the evil of stained, tainted literature, which has been lavishly imported into this once happy land'.[43]

Discussions surrounding Dublin's status as a degenerate city and a nucleus of depraved literature focused not only on the importation of 'vile novels' but on their production in Ireland. In 1925, Father R.S. Devane, one of the Free State's most vocal advocates of literary censorship, identified Dublin as a source of 'bad literature', suggesting that a 'blacklist' of books be compiled and asking: 'Why should not the Free State make a beginning with such a list of its own, and as a deterrent to the Dublin cloacal school, open it with the notorious volume of a well-known degenerate Irishman?'.[44] Devane referred to Joyce's *Ulysses*. The debates surrounding immoral literature in the 1920s encapsulated the shift in Irish popular constructions of degeneration; rather than identifying England as the sole locus of corruption, social purity campaigners now increasingly viewed Ireland and specifically Dublin as a source of literary filth. Appropriately for a text subversively positioned against Free State Catholicism, *At Swim-Two-Birds* is 'a novel about writing an Irish novel of the 1930s', and the immoral activities cautioned against in the Christian Brothers' publications and newspapers throughout the Free State are precisely those the narrator of *At Swim-Two-Birds* engages in.[45] Moreover, these behaviours—gambling, drinking, smoking, and discussing and producing immoral literature—are provocatively staged in specific urban locations.

'Horse-Play in the Streets'[46]

Free State characterisations of Dublin as a forum of iniquity are reflected in the narrator's city-centre activities, which are presented in a fragmentary manner, interspersed with excerpts from the narrator's own novel manuscript and other intra-narratives. One of the most flagrant provocations of Free State social purity is to be found in the novel's portrayal of Dublin's drinking culture. The narrator's first, and each of his subsequent, encounters with intoxicating beverages are in the city; Dublin's main thoroughfares provide the setting for his capitulation to the ravaging effects of spirituous liquors. The narrator recounts that he was 'walking through the Stephen's Green on a summer evening and conducting a conversation with a man called Kelly ... who was addicted to unclean expressions in ordinary conversation and spat continually'.[47] Kelly suggests that the students 'should drink a number of jars or pints of plain porter in Grogan's public house'.[48] Following Kelly's suggestion and entering the city-centre public house, the narrator strays from the lessons instilled by his evident Christian Brothers' education and begins his trajectory, as warned of by the Brothers, towards early death 'by a drunkard's fall, expiring ingloriously at the stair-bottom in a welter of blood and puke'.[49]

O'Brien exaggerates Christian Brothers' fears of the dangers of alcohol when the narrator announces that he immediately became 'addicted to brown stout in bottle, a drink which still remain[ed] the one that [he] prefer[ed] the most despite the painful and blinding fits of vomiting which a plurality of bottles ha[d] often induced in [him]'.[50] The narrator is quickly seduced by the drinking culture of the city centre, frequenting Parnell Street, Stephen's Green, Nassau Street, and Grogan's Public House. His initiation into the habitual consumption of intoxicating beverages takes on an additional meaning when we consider the likelihood of his not being from Dublin. The necessity of his living with his uncle in order to attend college indicates that he may originally be from elsewhere in Ireland, as O'Brien himself was. Yet Dublin and its inhabitants are responsible for his introduction and subsequent addiction to alcohol. This is particularly significant given the idealisation of rurality in nationalist rhetoric during the Free State, which went hand in hand with the demonisation of the urban. The Christian Brothers' emphasis on the importance of self-control in pleasures such as smoking and drinking has evidently been disregarded by the narrator, despite his awareness of the abstinence advice prescribed to him 'by an aged lay-brother'.[51] Ignoring the Brothers' warnings, the narrator succumbs to the pleasures of the city, surrendering himself to the 'pleasantly' impairing effects of alcohol on the mind and health.[52]

The 'blinding fits of vomiting'[53] brought on as a result of alcohol consumption are, notably, always tied in the narrative to specific city-centre locations. The narrator recalls, for example, that after 'drinking pints' he 'left a gallon of half-digested porter on the floor of a public house in Parnell Street'.[54] The narrator and his cohorts consume the city's moral poison in the form of intoxicants and, in turn, alter its topography after their own unscrupulous image. While the pious character Shaun in *Finnegans Wake*, an emblem of the post-independence puritanism associated with Éamon de Valera, redecorates Dublin by painting the town a nationalist hue, the narrator of *At Swim-Two-Birds* and his peers take on an ideologically opposing task: rejecting a Catholicised vision of national progress, they vandalise the city with their spit and vomit in a burlesque gesture of modern urban degeneracy. As they consume alcohol, they also deface the city by leaving 'half-digested porter' on the floors of public houses, 'fouling the flowerbeds' in St Stephen's Green, leaving 'large spits' on the pavement at Nassau Street, and expelling 'an unpleasant buff-coloured puke' near Lad Lane police station.[55]

Despite the clear textual anchoring of immoral activities to specific city-centre addresses, the narrator rarely moves with purpose or direction through the city. Somewhat like Bloom in the 'Lestrygonians' episode of *Ulysses*, he appears to be propelled through the city by forces outside of himself. Following a confrontation with his uncle, he recalls that he 'walked down to the centre of the town without adverting to [his] surroundings and without a predetermined destination', ending up at the Nelson Pillar on O'Connell Street.[56] Without paying attention to his surroundings, the narrator is drawn to a location in the North inner-city which was infamous in the 1930s for its squalor and tenements. The narrator notes that 'A slight fog, perforated by the constellation of the street-lamps, hung down on the roadway from the roofs of the houses' and, although 'there was no rain ... the streets were glistening'.[57] In contrast to the citizens who 'were moving in a quick active manner along the pavements', the narrator ambles slowly and aimlessly.[58]

Another immoral and specifically urban activity to which the narrator is disposed is loitering. He and his acquaintances walk 'slowly down to Grogan's' and later stand 'watching and talking at the corner' before eventually going to the 'moving pictures ... travelling to the centre of the city in the interior of a tramcar',[59] presumably to see an immoral production. The 'immoral pictures' complained of in the national press during the Free State feature in the narrative only as a token of urban degeneracy—their content is notably absent, comically not remembered or remarked upon by the narrator, thus making a mockery of the over-stated ill effects of cinema-going on youth morality. Surely, O'Brien intimates, if films were as immoral and corrupting as degeneration theorists claimed, the narrator would at least remember the picture he had been to see. His inability to remember the production, however, may also be because of drunkenness, as he reports that, following his addiction to brown stout in a bottle, events were often 'imperfectly recorded in [his] memory'.[60] The reader may guess that the young men went to see a cowboy film, as, at a later point in the novel, the character Shanahan, a creation of the narrator's character, remarks 'That place is a picturehouse now, of course ... plenty of the cowboy stuff in there. The Palace Cinema, Pearse Street. Oh, many a good hour I spent there too'.[61]

Strolling idly through the city and observing Dublin's citizens, the narrator is not wholly a *flâneur* or a *badaud*—the less glamorous gawker or bystander cousin of the flâneur—but a composite of both.[62] O'Brien exaggerates Free State fears of sexual immorality in Dublin when he has the narrator announce that the 'discovery and embracing of virgins'[63] was the purpose of his walk through the city with an acquaintance named Kelly, whom he encounters in a district which, the narrator is sure to point out, is 'frequented by the prostitute class'.[64] The city's topography becomes a playground for the two young men as they discover new opportunities for harassing married women, as well as discussing 'dog-racing, betting and offences against chastity' across the 'environs of Irishtown, Sandymount and Sydney Parade ... Haddington Road and the banks of the canal'.[65]

The city also fosters the narrator's interest in, and production of, immoral literature. At the novel's opening, the narrator explains that his bookshelf contained the works of 'Mr Joyce'.[66] Here, O'Brien does more than acknowledge the novel's place in a Joycean literary tradition—he emphasises its critical role in a censorious cultural climate in which Joyce's works were, if not outright banned, considered 'degenerate'.[67] Literary discussions which take place in the city centre are often focused on the value of immoral plots and characters. After the narrator realises he has reached the Nelson Pillar, a man named Kerrigan inconspicuously emerges from 'a side-street' and convinces him to walk in 'the Grafton Street direction', venturing to the house of a Dublin intellectual, Michael Byrne, for a discussion on modern literature.[68] Here, the narrator details the composition of his own novel and the immoral characters contained therein, such as Dermot Trellis, 'a philosopher and moralist' who is 'appalled by the spate of sexual and other crimes recorded in recent times in the newspapers',[69] but who nonetheless rapes and murders one of his own characters, Sheila Lamont. The narrator's composition of a novel which contains a novel in which there are 'no less than seven indecent assaults on young girls'[70] ridicules Free State suggestions that Dublin was a hub of immoral literature, mocking in particular the Christian Brothers' assertion that 'The character of the reading indulged in for some years past, is altogether responsible for the present depravity of literary tastes among the youth of our country'.[71]

As Liam Lanigan observes, such fears about the detrimental impact of metropolitan life on Irish purity were not always mocked, but also earnestly expressed in Free State literature, notably in the urban realism of Liam O'Flaherty's *The Informer*, where Dublin is 'a paranoiac space' of slums and grinding poverty.[72] Clearly, the literature of the Free State was varied in its response to theories of degeneration. O'Brien, however, refuses to take such concerns seriously, his unreservedly sardonic tone throughout the novel setting his rendering of the degenerate city apart from Free State and modernist predecessors.

While *Ulysses* is routinely cited as an example of literary high modernism and, as such, unequivocally holds the status of modernist, the question of *At Swim-Two-Birds*' position in the literary modernist canon is more complex, owing in part to its formative aesthetic innovations and its publication at a transitional moment in Irish, European, and transnational modernist history. Its multi-layered, fragmented, and reflexive narrative both extends modernism's disruption of novelistic structural conventions and anticipates postmodernist techniques of fragmentation and metafiction. As such, the novel has been interpreted according to the frameworks of both late modernism and postmodernism. For M. Keith Booker, *At Swim* 'resonates with important movements in postmodernist fiction', particularly in terms of its 'reflexive concern with its own status as fiction and of its own destabilization of traditional ontological boundaries'.[73] As Maebh Long has highlighted, the novel's fragmentary structure and aesthetic have been cited as evidence both of its postmodernist and late modernist features.[74] For Peter Childs, the fragmentary nature of *At Swim-Two-Birds*, as well as its publication date, situates it at the boundary between modernism and postmodernism, making it the ideal text to conclude his Routledge volume on modernism. *At Swim*, he argues, serves 'as one marker of the end of the Modernist Revolution', with O'Brien emerging at the end of a lineage of modernist formal and linguistic innovators who laid 'the foundations for postmodernist writers to undermine the very categories of "character", "representation", "reality", and even "literature"'.[75]

The text's undermining of 'reality' in particular is significant for thinking about the way the urban is represented. The reality of Dublin in the mid-1930s, infamous for the extreme poverty and disease which was concentrated in its slums, is far from the provocative image of the city—a playground of vice and indulgence for male undergraduates—presented by O'Brien in the book. Rather than communicating the socioeconomic realities of the city, O'Brien magnifies and exaggerates aspects of the urban most offensive to the moral architects of a Catholic post-independence Ireland. To borrow the words of another Free State urban novelist and late modernist writer, Samuel Beckett, O'Brien's rendering of a Dublin which is a haven of gambling, drinking, prostitution, and loitering appears 'carefully calculated to deprave the cultivated reader'.[76]

Like Joyce and other Free State urban writers such as O'Flaherty, O'Brien deploys the city and its attendant connotations in the service of a political aim—in *At Swim-Two-Birds*, an aim is evidently the subversion of

Catholic authority. Yet O'Brien differs from earlier Irish urban writers in that his Dublin is decidedly not the complex, multifaceted, and authentic city that colours the pages of *Ulysses*. It is a distinctly one-dimensional, parodically nefarious fictionalised zone whose cultural and literary associations with national decline are exploited by O'Brien in order to mock the discourses of social purity which dominated the cultural and moral landscape of the Free State. In its portrayal of the urban as a site of moral contagion, *At Swim-Two-Birds* deftly satirises Free State social purists' fears that Dublin offered too many opportunities for leisure and entertainment and was a distraction from Catholic values and the imperative of national moral purity. The comical vilification of the urban in *At Swim-Two-Birds* is predominantly a satiric response to the moral geographies which formed a core part of post-independence nationalist rhetoric, and O'Brien's wholly sardonic tone, coupled with the deliberate superficiality of the rendering of the urban, sets the novel apart from earlier literary representations of Dublin.

Notes

1. Flann O'Brien. *At Swim-Two-Birds*, p. 44.
2. Gerry Smyth notes that 'the major paradox informing any consideration of the city in cultural or social scientific terms … is that its representation as a fundamentally ambivalent and contradictory phenomenon has remained remarkably consistent … it was configured as either the primary location of 'civilization, culture and art', or as a place of corruption and constraints, a veritable hell on earth.' Gerry Smyth, 'The Right to the City: Re-presentations of Dublin in Contemporary Irish Fiction.' *Contemporary Irish Fiction: Themes, Tropes, Theories*, edited by Liam Harte and Michael Parker (Basingstoke: Palgrave Macmillan, 2000), p. 15.
3. Fragmentation is considered a key aspect of the literary modernist representation of the city. See Bart Keunen, 'Living with Fragments: World Making in Modernist City Literature.' *Modernism*, edited by Astradur Eysteinsson and Vivian Liska (Amsterdam: John Benjamins, 2007), pp. 271–277.
4. As Scarlett Baron notes, 'Joyce's meticulous attention to the physical, social, and geographical realities of Dublin inscribes *Ulysses* within the realistic tradition'. Scarlett Baron, 'Beginnings.' *The Cambridge Companion to Ulysses*, edited by Sean Latham (Cambridge: Cambridge University Press, 2014), p. 62. Maud Ellmann, too, writes that the challenging styles which confront the reader in *Ulysses* operate 'in the service of an intensified mimeticism that seeks to outdo—not, crucially, undo—narrative realism.

Ulysses might be an anti-realist novel, but that is only true insofar as it attempts to be more realistic than realism'. Maud Ellmann, 'Endings.' *The Cambridge Companion to Ulysses*, p. 113.

5. Frank Budgen, *James Joyce and the Making of Ulysses* (1934); reprint ed. (Bloomington, Indiana University Press, 1960), pp. 67–68. Quoted in Karen Lawrence, 'The Narrative Norm.' *The Odyssey of Style in Ulysses* (Princeton: Princeton University Press, 2001), p. 11.
6. Declan Kiberd. 'Gaelic Absurdism: *At Swim-Two-Birds*.' *Irish Classics* (Harvard University Press, 2001), pp. 500–520. Joseph Brooker. 'Estopped by Grand Playsaunce: Flann O'Brien's Post-Colonial Lore.' *Journal of Law and Society*, vol. 31, no. 1, March, 2004, p. 22.
7. Flann O'Brien. *At Swim-Two-Birds*. p. 6.
8. Ibid.
9. Ibid., p. 42.
10. Ibid., p. 93.
11. Ibid., p. 24.
12. Ibid., p. 7.
13. Gregory Dobbins. 'Flann O'Brien and the Politics of Idleness.' *Lazy Idle Schemers: Irish Modernism and the Cultural Politics of Idleness* (Dublin: Field Day, 2010), p. 196.
14. Flann O'Brien. *At Swim-Two-Birds*, p. 161.
15. Sandra L. MacAvoy. 'The Regulation of Sexuality in the Irish Free State, 1929–1935.' *Medicine, Disease and the State in Ireland, 1650–1940*, edited by Greta Jones and Elizabeth Malcolm (Cork: Cork University Press, 1999), pp. 253–267. Louise Ryan. *Gender, Identity and the Irish Press, 1922–1937: Embodying the Nation* (New York: Edward Mellen Press, 2002).
16. *The Irish Independent*, 22 January 1924. This is from an advertisement promoting the Irish Christian Brothers' publication, *Our Boys*.
17. Clair Wills, 'Fitness, Marriage, and the Crisis of the National Family.' *The Best Are Leaving: Emigration and Post-War Irish Culture* (Cambridge: Cambridge University Press, 2015), 23.
18. Patrick Sarsfield O'Hegarty, *The Victory of Sinn Féin* (Dublin: Talbot Press, 1924), 130–131.
19. Daire Keogh, 'Our Boys: The Christian Brothers and the formation of youth in the 'new Ireland' 1914–1944.' *History of Education*, vol. 44, no. 6 (2015), p. 709.
20. Keogh, p. 716.
21. Ibid.
22. Ibid.
23. Ciara Breathnach, 'Introduction: Ireland church, state and society 1900–1975.' *The History of the Family* vol. 13, no. 4 (2008): 335. See also K. Theodore Hoppen, 'Promised Lands: Ireland Since 1921.' *Ireland*

Since 1800: Conflict and Conformity. 2nd ed. (Abingdon: Routledge, 2013), pp. 185–291.
24. *The Evening Herald*, January 1924.
25. *The Evening Herald*, December 1923.
26. *The Irish Independent*, March 1931.
27. See Daniel Shea, 'A Rank Outsider: Gambling and Economic Rivalry in *Ulysses*.' *James Joyce Quarterly* vol. 48, no. 1 (2010): pp. 75–88.
28. 'A Great Social Evil: The Passion for Gambling." *Our Boys*, August 1927.
29. Ibid.
30. 'The Betting Craze.' *The Irish Independent*, October 1927. 'Dublin has the betting craze bad.' *The Evening Herald*, March 1930. 'Dublin's Gambling Plague.' *The Evening Herald*, September 1928.
31. 'Evils of Gambling: Dublin Police Ask For New Legislation.' *The Ulster Herald*, March 1928.
32. 'How Youth Fell: Justice and Betting Scandal: Dublin 'Gambling Hells'.' *The Irish Independent*, August 1930.
33. *The Irish Examiner*, May 1931.
34. 'Clergyman and Gambling.' *The Irish Independent*, March 1933.
35. Katherine Mullin, "Antitreating is about the size of it': James Joyce, Drink, and the Rounds System.' *Review of English Studies* vol. 64 (2013): pp. 311–328.
36. *Our Boys*, October 1931. *Our Boys*, April 1923. 'Self control.' *Our Boys*, May 1934.
37. *Our Boys*, December 1927.
38. Flann O'Brien. *At Swim-Two-Birds*. p. 18.
39. Mulcahy, Reverend W. 'Alcohol and Health – 1.' *The Christian Brothers: Higher Literary Reader* (Dublin: M.H. Gill and Son Limited, 1925), pp. 85–92.
40. *Our Boys*, November 1927.
41. *Our Boys*, July 1922.
42. *Our Boys*, November 1927.
43. Ibid.
44. R.S. Devane. 'Indecent Literature: Some Legal Remedies. Introductory: The Bishops' views.' *Irish Ecclesiastical Record* 25 (1925): p. 183.
45. Carol Taaffe, 'The Genesis of At Swim-Two-Birds.' *Ireland Through the Looking-glass: Flann O'Brien, Myles na gCopaleen and Irish Cultural Debate* (Cork: Cork University Press, 2008), p. 34.
46. Flann O'Brien. *At Swim-Two-Birds*. p. 55.
47. Ibid., p. 17.
48. Ibid.
49. Ibid.
50. Ibid., p. 19.
51. Ibid., p. 17.

52. Ibid., p. 22.
53. Ibid., p. 19.
54. Ibid.
55. Ibid., p. 20, 23, 39, 47.
56. Ibid., p. 93.
57. Ibid.
58. Ibid.
59. Ibid., p. 47.
60. Ibid., p. 36.
61. Ibid., p. 193.
62. See Charles Travis, 'From the Ruins of Time and Space: The Psychogeographical GIS of post-colonial Dublin in Flann O'Brien's *At Swim-Two-Birds* (1939).' *CITY: analysis of urban trends, culture, theory, policy, action* vol. 17, no. 2 (2013): pp. 209–233.
63. Ibid., p. 44.
64. Ibid.
65. Ibid.
66. Ibid., p. 7.
67. Devane, p. 183.
68. Flann O'Brien. *At Swim-Two-Birds*. p. 93.
69. Ibid., p. 31.
70. Ibid.
71. *Our Boys*, August 1927.
72. Liam Lanigan, 'Epilogue: Writing Dublin after Joyce.' *James Joyce, Urban Planning and Irish Modernism: Dublins of the Future* (Basingstoke: Palgrave Macmillan, 2014), p. 207.
73. M. Keith Booker, 'Flann O'Brien in the Twentieth Century.' *Flann O'Brien, Bakhtin and Menippean Satire* (New York: Syracuse University Press, 1995), p. 122.
74. Maebh Long, 'Introduction.' *Assembling Flann O'Brien* (London: Bloomsbury, 2014), p. 9.
75. Peter Childs, *Modernism* (London: Routledge, 2000), p. 207.
76. Samuel Beckett, *Murphy* (1938), edited by J.C.C. Mays (London: Faber & Faber, 2012), p. 75.

References

'A Great Social Evil: The Passion for Gambling.' *Our Boys*, August 4, 1927.

Baron, Scarlett. 'Beginnings.' In *The Cambridge Companion to Ulysses*, ed. Sean Latham, 51–66. Cambridge: Cambridge University Press, 2014.

Beckett, Samuel. *Murphy (1938)*. London: Faber & Faber, 2012.

Booker, M. Keith. *Flann O'Brien, Bakhtin and Menippean Satire*. New York: Syracuse University Press, 1995.

Breathnach, Ciara. 'Introduction: Ireland Church, State and Society 1900–1975.' *The History of the Family* 13, no. 4 (2008): 333–339.

Brooker, Joseph. 'Estopped by Grand Playsaunce: Flann O'Brien's Post-Colonial Lore.' *Journal of Law and Society*, 31, no. 1 (March, 2004): 15–37.

Budgen, Frank. *James Joyce and the Making of Ulysses (1934)*; reprint ed., Bloomington: Indiana University Press, 1960, 67–68.

Childs, Peter. *Modernism*. London: Routledge, 2000.

'Clergyman and Gambling.' *The Irish Independent*, Saturday March 25, 1933, p. 7.

Cruise O'Brien, Conor. *Ancestral Voices: Religion and Nationalism in Ireland*. Dublin: Poolbeg, 1994.

'Degenerate Dublin.' *The Irish Independent*, Thursday May 22, 1924, p. 8.

Devane, R.S. 'Indecent Literature: Some Legal Remedies. Introductory: The Bishops' Views.' *Irish Ecclesiastical Record* 25 (1925): 182–204.

Dobbins, Gregory. *Lazy Idle Schemers: Irish Modernism and the Cultural Politics of Idleness*. Dublin: Field Day, 2010.

'Dublin has the Betting Craze Bad.' *The Evening Herald*, Saturday March 29, 1930, p. 4.

'Dublin's Gambling Plague.' *The Evening Herald*, Thursday September 6, 1928, p. 6.

'Editor's Notes.' *Our Boys*, 8.11, July 5, 1922.

Ellmann, Maud. 'Endings.' In *The Cambridge Companion to Ulysses*, ed. Sean Latham, 95–128. Cambridge: Cambridge University Press, 2014.

'Evils of Gambling: Dublin Police Ask for New Legislation.' *The Ulster Herald*, Saturday March 10, 1928, p. 6.

Hoppen, K. Theodore. *Ireland Since 1800: Conflict and Conformity*. 2nd ed. Abingdon: Routledge, 2013.

'How Youth Fell: Justice and Betting Scandal: Dublin "Gambling Hells".' *The Irish Independent*, Wednesday August 13, 1930, p. 4.

Keogh, Daire. 'Our Boys: The Christian Brothers and the Formation of Youth in the "New Ireland" 1914–1944.' *History of Education* 44, no. 6 (2015): 700–716.

Keunen, Bart. 'Living with Fragments: World Making in Modernist City Literature.' In *Modernism*, ed. Astradur Eysteinsson and Vivian Liska, 271–277. Amsterdam: John Benjamins, 2007.

Kiberd, Declan. 'Gaelic Absurdism: *At Swim-Two-Birds*.' In *Irish Classics*, 500–520. Harvard University Press, 2001.

Lanigan, Liam. *Dublins of the Future: James Joyce, Urban Planning, and Irish Modernism*. Basingstoke: Palgrave Macmillan, 2014.

Lawrence, Karen. *The Odyssey of Style in Ulysses*. Princeton: Princeton University Press, 2001.

Long, Maebh. *Assembling Flann O'Brien*. London: Bloomsbury, 2014.

MacAvoy, Sandra L. 'The Regulation of Sexuality in the Irish Free State, 1929–1935.' In *Medicine, Disease and the State in Ireland, 1650–1940*, ed. Greta Jones and Elizabeth Malcolm, 253–267. Cork: Cork University Press, 1999.

Mulcahy, Reverend W. 'Alcohol and Health–1.' In *The Christian Brothers: Higher Literary Reader*. Dublin: M.H. Gill and Son Limited, 1925.

Mullin, Katherine. '"Antitreating is About the Size of It": James Joyce, Drink, and the Rounds System.' *Review of English Studies* 64 (2013): 311–328.

O'Brien, Flann. *At Swim-Two-Birds (1939)*. London: Everyman, 2007.

O'Hegarty, Patrick Sarsfield. *The Victory of Sinn Féin*. Dublin: Talbot Press, 1924.

Our Boys, April, 1923.

———, October, 1931.

Ryan, Louise. *Gender, Identity and the Irish Press, 1922–1937: Embodying the Nation*. New York: Edward Mellen Press, 2002.

'Self Control.' *Our Boys*, May 1934.

Shea, Daniel. 'A Rank Outsider: Gambling and Economic Rivalry in *Ulysses*.' *James Joyce Quarterly* 48, no. 1 (2010): 75–88.

Smyth, Gerry. 'The Right to the City: Re-presentations of Dublin in Contemporary Irish Fiction.' In *Contemporary Irish Fiction: Themes, Tropes, Theories*, ed. Liam Harte and Michael Parker, 13–35. Basingstoke: Palgrave Macmillan, 2000.

Taaffe, Carol. *Ireland Through the Looking-Glass: Flann O'Brien, Myles na gCopaleen and Irish Cultural Debate*. Cork: Cork University Press, 2008.

'The Betting Craze.' *The Irish Independent*, Monday October 17, 1927, p. 5.

The Evening Herald, Thursday December 13, 1923, p. 2.

———, Thursday January 31, 1924.

The Irish Examiner, Wednesday May 20, 1931, p. 8.

The Irish Independent, January 20, 1924.

Travis, Charles. 'From the Ruins of Time and Space: The Psychogeographical GIS of Post-Colonial Dublin in Flann O'Brien's *At Swim-Two-Birds* (1939).' *CITY: Analysis of Urban Trends, Culture, Theory, Policy, Action* 17, no. 2 (2013): 209–233.

'Well Done, Dublin!.' *Our Boys*, 14.8, December 9, 1927.

'What Irish Hearts Can Do.' *Our Boys*, 14.6, November 10, 1927.

Wills, Clair. *The Best Are Leaving: Emigration and Post-War Irish Culture*. Cambridge: Cambridge University Press, 2015.

CHAPTER 8

Putting the 'Urban' into 'Disturbance': Kevin Barry's *City of Bohane* and the Irish Urban Gothic

Martyn Colebrook

The city, our great modern form, is soft, amenable to a dazzling and libidinous variety of lives, dreams, interpretations. But the very plastic qualities which make the city a great liberator of human identity also cause it to be especially vulnerable to psychosis and totalitarian nightmare. (Raban 1974, 8)

Kevin Barry's strangely dislocating debut novel is a fictional depiction of a remote dystopian city located on the west coast of Ireland—said to be inspired by Limerick and Cork.[1] Barry's novel fits into the context of the Irish Gothic with its 'landscape of ruins and rack-rented estates' (Sage, 138), as well as the tradition of the wider Gothic and, arguably, Celtic Speculative Fiction, oriented around such literary precedents as Alasdair Gray's *Lanark: A Life in Four Books* (1981), Hal Duncan's *Vellum* (2006) and, more globally, the fiction of China Mieville, rooted through a Lovecraftian fixation on the intersections between language and geography.[2] The chapter will consider the depiction of the urban in relation to its construction using the language of the urban Gothic and place *City of*

M. Colebrook (✉)
Independent Scholar, London, UK

M. Beville, D. Flynn (eds.), *Irish Urban Fictions*, Literary Urban Studies, https://doi.org/10.1007/978-3-319-98322-6_8

Bohane within the wider discussion of the Irish Urban Gothic, using critical frameworks established by Catherine Spooner and Linda Dryden. The consideration of the urban in relation to the Irish Gothic stems initially from an identifiable trope in contemporary novelists such as Stuart Neville, Ken Bruen and, to a lesser extent, Patrick McCabe (to name but a few) whose interrelation of the Gothic and the Noir creates the opportunity for depictions of cityscapes and townscapes which appear to exist as living entities alongside the inhabitants of these environments. It is as though there is a tension between the question of whether the inhabitants shape the Gothic characteristics of the urban or the urban environment shapes the Gothic tendencies of its inhabitants.

Given the tensions outlined above and the manner of the Gothic to infiltrate many different artistic forms and environments, it is unsurprising that critics have identified artistic influences beyond the literary lineage noted in the previous paragraph. As Hamill notes, 'he evokes the graphic language of certain master comic book artists from Will Eisner and Harvey Kurtzman all the way to Frank Miller, of *Sin City*. That style is noirish: cinematic long shots, medium shots, close-ups, played against the deep, rich blacks of ominous night. Like a great movie director, Barry always pays careful attention to the way his characters dress'.[3] Among the comments providing the critical lauding upon *City of Bohane*'s publication, most relevant to this chapter are the suggestions that 'the landscape is part-Dickens, part-Escheresque computer game, and the same goes for its plot'[4] and *The Sunday Times*' statement: '[t]his slangy, plosive-packed prose is what makes the book a success … [i]t is a plus point that the dystopia bears no allegorical weight, thriving purely as an imaginary realm to be taken at face value'.[5] These references to Dickens and Escher highlight the position of the novel's structure as a vital pivot in relation to the wider text. The writer thus presents a case for considering both the structure and the language and, potentially, their interrelationship.

Considering the city and language, it is notable that Barry suggests the hardest element of the writing involved: recasting the English language to convincingly describe the life of a malevolent west of Ireland city in the middle of the twenty-first century. The novel is a weird retro-fitted future-Western, with lots of gratuitous swearing, hideous violence, perverse sex and powerful opiates, but above all it's a projection of what the street talk might sound like among hipster hoodlums in such a city in the 2050s.[6]

Language in Bohane functions not only as a tool for communication but also for preservation, recording and documenting the city's evolutions and revolutions. Barry references the 'street talk' he imitates, and this emphasises the significance of the urban Gothic's relationship with the post-industrial. The significance of the Gothic in relation to 'post-industrial space' is suggested by Linda Dryden who observes that: [t]he city itself is figured as monstrous in its geographical and social divisions. The duality that some of its citizens exhibit is also manifest in the oppositions of light and dark, atmospheres of airiness and stifling oppression.[7] In this depiction, the city becomes 'corrupt, diseased',[8] a site of anxiety and decay. Additionally, there is evidence of other tenets of the urban Gothic, such as the intermingling of class-focused violence, rough work, taverns and brothels which act as sites of transgression, opportunities for social meetings and covers for nefarious business exchanges within the city.

Barry has already explained in his interview that Bohane constitutes a multi-layered entity composed of the past, the present and the future potential. As such, the Gothic resonates as being of the city as opposed to in the city. It is this emphasis on memories and historical associations which aligns Bohane with the Urban Gothic. This emphasis runs through the heart of the plot that explores the fractures and tribalism between Logan Hartnett and his family, whose legacy of criminal activities has made for a scarred and troubled community, and Harnett's nemesis, Gant Broderick, whose return threatens a new period of future instability. A notable constant amongst the reviews identifies an almost consciously Lovecraftian technique at work within Barry's novel, whereby the 'plot', a straightforward return of a former hero and a bloody conclusion to an inter-city conflict, functions as the window-dressing for Barry's real love; in this instance, the languorous linguistic play and atmosphere are the key desiderata underpinning Barry's novel. The process is akin to the strategies devised by pulp novelists which produces the strange and unsettling sensation where you are reading 'a pulp bricolage where texts concatenate out of scattered scraps, in what looks like a deliberate undermining of "plot"'.[9] This is not to say that Barry abandons the formal narrative and structure for a more experimental technique but to suggest that the area of Smoketown itself, a labyrinthine testament to Gothic cityscapes and tropes, is a bricolage itself of slums and backstreets which sprawls onto the vast expanse of the Big Nothin', an eerie bog which surrounds the city. The city's estranging and unsettling structure for the unwary is a reflection of the technique identified by Mieville. The use of space in Mieville, and the wider Gothic, functions as a mechanism by which 'human experience'[10] or 'mood'[11] can be commented on, whether that be the terror of what lies hidden or that which is lurking just beyond the limits of human perception. The city functions as a liminal space within an urban setting, a veritable doppelganger to the surrounding area. The city represents a distinctive space which functions as a site of Gothic decay, disorder and dissent, a sequence of heterotopic nodes around which individual communities gather and merge, as well as a layered construction of historical entities in a state of continuous flux. To this effect, Dryden observes that '[i]ssues of duality – split personalities, physical transformations, mistaken identities, doppelgangers – were found to be manifested in the social, geographical and architectural schisms of the modern city'.[12] Ultimately our attention to the terrain of a Gothicised city serves to emphasise the significance of the built environment for the ways in which it functions as a metaphor for socio-psychological crisis.

Indeed, the title itself, *City of Bohane*, draws attention to the terrain but also inverts the expected grammatical structure and order by describing the location as being 'of the city' rather than the city being defined by the location. Barry's assertion of a 'retro-fitted future-Western' indicates the temporal juxtaposition he achieves, taking the past and elegantly projecting it onto an imagined future in a manner which would usually be associated with the conventions of science fiction, but exists in a book where 'there is no science and no technology'.[13] This absence of the two significant components for science fiction points to Barry's position as another contemporary Irish Gothic novelist who straddles and breaches the boundaries between genres whilst demonstrating his admiration for 'the work that's being done in what people still insist on calling "the genres"', explaining further that 'it's critical now that writers work to break down the false divides and structures'.[14] Those 'false divides and structures' recur through Barry's views on the process of writing when he speaks of it being 'very close to dreaming' and how writing a few sentences first thing in the morning 'is a wonderful time to explore the true depths'.[15] His reference to Don DeLillo's work demonstrates an awareness of the techniques which inform his prose; the city's status as a work of imagination with no allegorical correlation permits his outlandish depictions and unashamed acknowledgement of his literary antecedents—indeed, this novel remains one of those where the novelist can expand the limits of linguistic play and remain strikingly close to the fantastical Gothic, pulpish tone he sets out to achieve.

Giving thought to this Gothic, pulpish tone, the first chapter of *City of Bohane* is tellingly titled 'The Nature of the Disturbance' and emphasises the existence of a threat to the stability of the city. The reader is invited and introduced to Bohane with these opening lines:

Whatever's wrong with us is coming in off that river. No argument: the taint of badness on the city's air is a taint off that river. This is the Bohane river we're talking about. A blackwater surge, malevolent, it roars in off the Big Nothin' wastes and the city was spawned by it and was named for it: city of Bohane.[16]

Barry uses the river as the driving motivation and reason for the 'taint of badness'[17] within the city, and it is easy to interpret the features and realise that Bohane contains all the trappings of a port. The language establishes the sensations of a dark, dangerous environment, the *mise-en-scène* of the urban Gothic poisoned by the 'taint' of a 'malevolent' river, a 'blackwater surge' which powerfully 'roars' into the city, bringing with it the toxic 'wastes' from the Big Nothin'.[18] That the 'city was spawned by it'[19] indicates that this is not a pleasant or conventional birth, instead there is an alien quality to the city and it remains an unwanted and undesired product. The port offers the perfect opportunities for unregulated deposits and trades, unseen meetings as well as being populated by those willing to indulge the vices and desires of the deviants and fantasists who must seek their pleasures out of sight. For the residents, the river represents 'whatever's wrong with us', an unidentifiable, malignant undercurrent further embellishing the alien qualities of the city and the unsettling foreignness which one identifies with the urban Gothic.

The subtext in these early chapters involves a threat posed to the urban order by an individual from the outside (later identified as Gant Broderick), paralleling the structure of the crime genre's locked-room mystery in which the threat to the domestic is posed by an individual from without. This is also the classic motif from the Western in which the stranger comes 'into town' in order to challenge or disrupt the existing social order. As he 'walked the docks', there is a suggestion that the individual is familiar with the geography of Bohane—a lack of knowledge in an area such as this would not lead logically to one traversing the area 'past midnight', nor would it allow for the confident patrolling which this individual appears to be undertaking. The aroma of 'sweet badness'[20] which he inhales further establishes an atmosphere redolent with criminality and vice, not an oxymoron but a clue as to the delight that is taken in such activities and behaviour. This is a further indication of the dark and deviant activities which rupture the fabric of the urban and render it Gothic.

The 'evenness to his footfall' which the previous quotation highlighted reinforces this familiarity and confidence, an absence of panic or uncertainty in his step which leads to the 'calm rhythm of leather on stone'.[21] That the dockside lamps 'burned' suggests the absence of electricity and a reliance on traditional industry with the 'green haze' functioning as an indicator of pollution. The final simile, 'the light of a sad dream' indicates that this location has been forgotten in time; the legacy referring back to the manner of Bohane's creation by a river transporting waste.[22] The phrase suggests a heavy-hearted acceptance of the area's run-down status where the dreams of the individuals who populate and work in this location have been reduced to an equally hopeless status. Whilst Hartnett is a long resident of Bohane, the language which infuses this sequence further reinforces the Gothic qualities of memory and association that inform the city's depiction.

As Hartnett continues his walk, the narrative intensifies the tense anticipation which accompanies him:

> The water's roar for Hartnett was as the rushing of his own blood and as he passed the merchant yards the guard dogs strung out a sequence of howls all along the front. See the dogs their hackles heaped, their yellow eyes livid. We could tell he was coming by the howling of the dogs.[23]

As we discover the name, Hartnett, the tension is heightened for the rushing water of Bohane is compared with the adrenaline that he is experiencing upon his return. The guard dogs reacting by giving 'out a sequence of howls' simultaneously emphasises the alien and foreign qualities associated with Hartnett.[24] The dogs either recognise him as a source of danger, symptomatic of the Gothic ambience of Bohane, or they cannot recognise him and thus react with wariness and fear. That the 'hackles heaped' is aggression and indicates the threat Hartnett poses to the security of the site they have been assigned to defend, their 'livid' eyes suggest a wild anger.[25] The narrator can 'tell he was coming' which suggests this is an unwelcome appearance from a dangerous or malignant force—an individual who has disturbed the fabric of the city and is recognised for the threat he poses.[26] The image of an individual who presents such a threat to the fabric of the city—albeit a city whose menacing presence is conveyed through the language from which it is constructed—contributes to the Gothic atmosphere Barry sustains throughout the novel.

As the menacing Hartnett continues his journey, he moves away from the docks and begins to enter the city: '[h]e cut off from the dockside and walked on into the Back Trace, the infamous Bohane Trace, most evil labyrinth, an unknowable web of streets. [...] It was a pair of Portuguese boots that slapped his footfall and the stress that fell, the emphasis, was money'.[27] The reader is now confronted with their first depiction of the city itself as separate from the docks. The 'most evil labyrinth' highlights both the structural complexities of the urban landscape which can entrap and ensnare the unwary or unknowing and the dark Gothic qualities which inform and underpin.[28]

These structural complexities mean that the streets are 'unknowable', suggesting that even to the individuals who are familiar and local to Bohane, the Back Trace remains an area of potential confusion and danger.[29] The 'web' structure indicates its propensity for trapping people, entangling them in such a manner as they are helpless until the inevitable danger consumes them. Hartnett's status within the city is hinted at with the reference to the 'emphasis' of 'money' which accompanies his gait.[30] Although Bohane has several opulent places (according to the precis on the cover), Smoketown is the area associated with poverty. The juxtaposition of the wealthy with the poor suggests his wealth has been accumulated from the proceeds of criminal or illicit activity, or exploitative practices which have bred resentment against him among the residents of this area. His demonstration of this through his choice of clothing and the confidence it provides ensures Hartnett is distinctive in this city.

More is revealed about the Trace as Hartnett continues his journey through the city: 'Dank little squares of the Trace opened out suddenly, like gasps, and Logan passed through. [...] He walked the Arab tangle of alleyways and wynds that make up the Trace and there was the slap, the lift, the slap, the lift of the Portuguese leather on the backstreet stones'.[31] The deplete and decrepit nature of the city is revealed here as we encounter the dank little squares. Cramped and squalid, they seemingly reveal themselves as Hartnett approaches. His power and influence means he has access to even the most stubborn or resistant of spaces. That they open 'like gasps' surreptitiously suggests the mild shock of the city at Hartnett's return, whilst 'Arab tangle' indicates a claustrophobic and probably densely populated urban area where buildings are squashed uncomfortably against one another.[32] By contrast, as Hartnett moves through the city, the narrator reveals that '[h]e feared not the shadows, he knew the fibres of the place, he knew every twist and lilt of it'.[33] There is a comparable absence of concern, given his familiarity with the locale and his confidence when striding through his domain—the 'fibres' of the city give it the personification of a fabric which has been stitched together rather than the loose and unstructured load that was dumped earlier from the river. Arguably, the reading from this perspective suggests that Hartnett's interpretation of the city is informed by his status: it is a more ordered and comfortable environment for him to operate in despite the latent threats which present themselves at every turn.

During Hartnett's time in Bohane, the rivalry between the Hartnett Fancy and the Cusacks was characteristically pronounced. The tradition in Bohane is that 'families from the Northside Rises will butt heads against families from the Back Trace', thus highlighting the geography of the city as subject to the frequent conflicts between the crime families characterises the depiction of a Gothic urban space riddled with divisions and underpinned by illicit economies and violence.[34] Reflecting further, it is revealed that 'his was the most ferocious power in the city that year. But here were the Cusacks building strength and gumption on the Rises'.[35] As Hartnett moves through the city, he sees Jenni Ching. Noting their status within the city's population, it is indicated that the Smoketown area of Bohane is reflected in Ching's status as one of those:

> old Smoketown stock. Smoketown was hoors, herb, fetish parlours, grog pits, needle alleys, dream salons and Chinese restaurants. Smoketown was the other side of the footbridge from the Back Trace, yonder across the Bohane river, and it was the Hartnett Fancy had the runnings of Smoketown also. But the Cusacks were shaping for it.[36]

There is a sense of generational longevity that accompanies this family, and Jenni's status as a *confidante* to Hartnett indicates the close relationship between the two. The forging of such an alliance is borne out of Jenni's service as a prostitute to Hartnett. The depiction of Smoketown is that of a run-down inner city where illegal activities form the social and economic dynamic which sustains the urban fabric of the area. Urban dereliction and decay can be seen through the 'hoors, herb, fetish parlours, grog pits, needle alleys, dream salons' and the socially transgressive behaviour which occurs within these sites.[37] For Catherine Spooner, writing about the contemporary Gothic, characteristics and concerns of this genre are the 'legacies of the past and its burdens on the present; the radically provisional or divided nature of the self; the construction of peoples or individuals as monstrous or "other"; the preoccupation with bodies that are modified, grotesque or diseased'.[38]

Hartnett is described as having a 'mouth of teeth on him like a vandalised graveyard' and the deformed Albino who observes his passage across Bohane is just another indication of the physical deformity and decay which imposes itself upon the population of the city.[39] This fascination with physical deformity, psychological or mental instability concurs with Dryden, who asserts that 'Gothic fiction is often a literature of transformations where identity is unstable'.[40] Identifying the emphasis on 'transformations', there is a focus on the historical and its influence on the contemporary, as well as the degeneration and breakdown of the body and the mind.

The geography of Smoketown is rendered in further Gothic detail as Hartnett moves through the city:

> Smoketown laid out its grogshops, its noodle joints, its tickle-foot parlours. Its dank shebeens and fetish studios. Its shooting galleries, hoor stables, bookmakers. All crowded in on each other in the lean-to streets. The tottering old chimneys were stacked in great deranged happiness against the morning sky.[41]

Smoketown is further depicted as a site of cramped streets and considerable urban decay—'shebeen' representing an illicit drinking den where illegal alcohol can be sold without a licence.[42] The economies of the body—drugs and sexual activity—are sharply consistent with the poverty of the area. The 'deranged happiness' of the old chimneys suggests a madness is inherent in the buildings and their structures, they are symbolic of industry and an architectural legacy from the past.[43]

Amongst this inherent madness, there is a persistent and latent threat of violence which contributes powerfully to the atmospheric nature of Barry's writing and depiction of the city. As the narrator explains: '[i]n a small city so homicidal you needed to watch out on all sides. He moved on through the gloom of the Back Trace. The streets of old tenements are tight, steep-sided, ill-lit, and the high bluffs of the city give the Trace a closed-in feel'.[44] The nature of the city as 'homicidal' suggests that it is the urban landscape itself which contributes to the murder and violence, as though the Gothic conditions influence the behaviour of the citizens and Barry continues to depict the city as a living character within the text.[45] The 'gloom' and 'old tenements' with their claustrophobic 'tight, steep-sided, ill-lit' structures give Bohane an irredeemably murderous feel, an intimidating and tense atmosphere in which the reader senses that mayhem, scores to be settled and violence may erupt (and irrupt) at any given moment.[46] The 'closed-in feel' intensifies this atmosphere: Bohane does not acknowledge the external world and remains insular, backward looking, while resentment and revolution breed and build on the straitjacketed streets and behind the doors of the taverns and tenements.[47]

Given the insularity of the Gothic Bohane, it's unsurprising that Chris Baldick identifies this atmosphere as one of the key characteristics of a Gothic text. Such a text should comprise a 'fearful sense of inheritance in time with a claustrophobic sense of enclosure in space, these two dimensions reinforcing one another to produce an impression of sickening descent into disintegration'.[48] This confrontation with and assertion of fears and anxiety from the past reinforces that 'Gothic is about breakdown, about terror, about the collapse of territory, structure, order, authority'.[49] A Gothic text flourishes in spaces which imprison or restrict the protagonists' efforts to move or exist comfortably and that the combination of both circumstances creates the feeling of 'disintegration' or 'fragmentation'. As if to impart the truly isolated and desolate nature of Bohane in relation to the rest of Ireland, the narrator reveals:

> Our city is built along a run of these bluffs that bank and canyon the Bohane river. The streets tumble down to the river, it is black and swift-moving rush at the base of almost every street, as black as the bog waters that feed it, and a couple of miles downstream the river rounds the last of the bluffs and there enters the murmurous ocean. The ocean is not directly seen from the city, but at all times there is the ozone rumour of its proximity, a rasp on the air, like a hoarseness. It is all of it as bleak as only the West of Ireland can be.[50]

That the streets 'tumble' on their way to the river suggests a chaotic lack of coordination and control, it encircles and surrounds the city, thus cutting it off from the remainder of the country.[51] Naming the expanse of landscape around Bohane 'Big Nothin' is telling, reflecting the absence of habitable environment or the presence of civilisation. The description of the possibility that the ocean may exist as being 'ozone rumours of its proximity' suggests Bohane is isolated and uses said rumours as a protective layer from the respective sunlight of surrounding country.[52]

Bearing in mind the wider functions of rumours, Hartnett enters a tavern in search of information and he is greeted by the barman, Tommie. The conversation that ensues is terse, laced with a hint of paranoia which is inculcated by the unspoken threat of reprisal if one is identified as an informant or worse:

> Tommie could not hide from his face the feeling that was current in the room, the leanings and nuance of the talk that swirled there. Logan used him always as a gauge for the city's mood. Bohane could be a tricky read. It has the name of an insular and contrary place, and certainly we are given to bouts of rage and hilarity, which makes us unpredictable.[53]

The use of 'current' and 'swirled' plays on references to the present time but also potentially alluding to both the 'electric' nature of the feelings in the atmosphere and the predictable flow of the conversation which is occurring in the tavern.[54] The 'leanings' and 'nuance' reflect the inner workings of Bohane—where what is not said may be equally as important (if not more important) than that which is articulated by the locals.[55] That Hartnett sees Tommie as a 'gauge' for the tone and mood in the city establishes the tavern as a site for the exchange of information, rumour, and suspicion.[56] The ubiquitous figure of the barman characterises the manner in which such information is divested and communicated: verbal, untraceable and thus reducing the chances of unintended punishment should implicated parties find out the source. The 'insular' and 'contrary' nature of the city demonstrates its engagement with duplicity and doubling, a schizophrenic mindset which is as split as its population.[57] The duality between 'rage' and 'hilarity' exemplifies this bipolarity, suggesting that the city itself may be characterised as a double within the narrative—the present being haunted and informed by its histories.[58] The atavistic nature of the Gothic and the manner in which terror and violence are inherited through generations are apparent in Bohane when one considers the close-knit families which are so prominent in their control of the different sections of the urban structure. These states provide a firm connection with the historical origins of the Gothic, which enables 'the construction of a convenient set of dualisms, primitive versus civilized, barbarism versus culture'.[59] These dualisms feel ingrained in the structure of Bohane—both at an individual and a geographical level.

Considering this principle of the structure of Bohane, the origins of the Hartnett family's relationship with the city establish a clear relationship between the Gothic and the city:

> The Hartnett seat was a Beauvista Gothical, a gaunt and lumbering old pile, all elbows and chimneys. Its thin, tall windows were leaded and reproachful, its gable ivied, the brickwork sharply pointed and with a honeyish tone that emerged fully now against the blue of a late morning in October. It sat plumb on a line of po-faced old manses that made a leafy avenue up top of the Beauvista bluff. The Bohane Dacency had built their Beauvista residences to face away from the city – though the money that built them had been bled from it – but Logan Hartnett and his wife were Trace-born, the pair of them, and they kept a rooftop garden on a terrace shaded by the chimney stacks, and it was oriented to look back across the great bowl of the city, as though in nostalgia for it.[60]

The Hartnetts are thus firmly characterised as of wealth and elevated social status—another key component of the Gothic. The family house and its 'gaunt' and 'lumbering' status as an 'old pile' suggest an unwieldy and large structure symbolising wealth and power.[61] The 'reproachful' windows are stern and express disapproval over those who may use them as a method to see into the Hartnett's family life.[62] The 'ivied' gable also suggests the creeping passage of time that has seen the family establish and assert their rule over Bohane, while the toxic nature of the berry associated with the ivy is potentially a reference to the dangers associated with attempting to overthrow or overwhelm the Hartnett empire as well as the hidden dangers associated with threatening the family.[63] That it sits 'plumb' in the middle of a 'leafy avenue' of other houses indicates the prominence of the position, as though a promontory looking down on the residents of Smoketown.[64] For Len Platt, the fall of the aristocracy 'triggered astonishing literary responses. In literary works, aristocrats were transformed into warrior heroes, Scotland Yard detectives, swashbucklers, diseased degenerates, and Gothic monsters', and Barry engages with but ultimately subverts this concept through Hartnett.[65] Despite the criminal activities which the family engage in, it is apparent that the family is seen as defending their territory and the comparative stability of Bohane against Gant Broderick upon his return to the city. The element of subversions comes from the manner in which the Hartnetts have established their elevated status, given that they are both born in the much worse area of the 'Trace'. They have forced their way into upper social circles through criminality and wealth acquisition from other people's production, highlighting the Gothic nature of Bohane as a city.

By comparison with the sturdy and robust home of the Hartnetts, Smoketown is presented as ramshackle, lacking in permanence or security:

> This Smoketown you take one brick from the pile and the whole heap'd come tumbledown. Smoketown it don't even make a square mile in size: a tight, small, squashed-up place, hard-pressed its airways, its troubled lungs, and the air had an oily feel in night.[66]

The image of a city in a state of collapse after one brick is removed suggests not only the provisional and precarious state of its industries and structures but also reflects that of its population—low-wage workers for whom the nature of their existences involves living from payday to payday and whose prospects are dependent on the whim of the employer. The lack of size has been discussed before, but it is the idea of 'hard-pressed [...] airways' and 'troubled lungs' which humanises the city as though there exists an interrelationship between the population and the landscape which comes to embody them.[67] The 'troubled lungs' add to this image of a cancerous or emphysemic body struggling to breathe easily after years of work in heavily industrial areas.[68] The 'oily feel' to the air emphasises how its cloyingly oleaginous sensation clings to the skin long after leaving the workhouse, a bleak atmosphere where one can never quite escape the last vestiges of their work.[69] Despite this sense of hopelessness, it becomes apparent that even Smoketown has its hierarchies:

> now the dune end of Smoketown is the cheaper line. There you will find a very low class of customer. The worst of the slagshops and the most insalubrious needle galleries are out there. Weird atmosphere on account of the system of dunes that rise just beyond and give the name to it. Haunted by ferocious pikeys it is.[70]

The depiction Barry gives is one of a ghetto or site of extreme social disenfranchisement within the city itself. This extreme social disenfranchisement within the urban site is the locale of Lovecraft and the worst excesses of transgression and decay within the city—lawlessness is rife and there is an inescapable sense that any sense of redemption is well-nigh impossible.[71] The Rises are developed further in their structure: '[t]he avenues of the Rises were broad, treeless, broken, and laid out to a vaguely Soviet pattern. The cement facades of the flatblocks were cracked from decades of freeze-and-thaw'.[72] This is the legacy of a post-industrial society where poverty and the totality of disenfranchisement are starkly apparently in their impact upon the cityscape. The idea of 'freeze-and-thaw' is suggestive of the cyclical nature of the city and its inhabitants—social control followed by total release.[73]

Thinking of the cyclical within the urban environment, as the feud between the Harnetts and the Cusacks escalates following a 'reefing' (knifing) in Smoketown, a notice is issued. The violence-drenched atmosphere intensifies and: '[t]he city's mood was a blend of fear and titillation. There was going to be an almighty collision, and a small world shudders when giants collide'.[74] The juxtaposition of 'fear' and 'titillation' emphasises the nature of the Gothic as a voyeuristic thrill that is gained by the sight of the forbidden, an erotic fascination with the violent or the transgressive.[75] As the reprisals continue and increase, 'the nerves of the city were ripped' and 'the city simmered now with bitterness, rage and threat'.[76] Like the entropic cities of Mieville and Michael Moorcock, Bohane is replete with a latent hostility and personal dread which threatens its population.

Significantly, Barry invokes a final Irish Gothic trope before the major conflict within the novel occurs, and it seems a fitting point at which to demonstrate a final node upon which the Gothic is located within *City of Bohane*. As Ol' Boy Mannion, Bohane's chief fixer, reflects when he visits the Big Nothin', the vast expanse which surrounds Bohane: '[t]he bog's occult nature had been interfered with, its body left scarred, its wounds open, and might this also be a source of the Bohane taint?'[77] That the source of Bohane's woes may be due to the corruption of a bog is steeped in Irish culture. The 'occult nature' of the bog suggests the supernatural, the ghostly, but Barry goes further and locates the bog as a scarred body, violated and wounded.[78] That the body is almost 'buried' enhances the status of the city as founded on the Gothic, and the manner in which this body 'taints' and influences from the dead is a striking image on which to base the structure.[79] This proverbial haunting of the city from beyond brings the reader full circle, emphasising the dynamics and the principles upon which Bohane is forged—a city with a dangerous and alien nature, brought to birth by the dumping of waste.

Notes

1. In an interview with *The Paris Review*, Barry explains that while Bohane is an invented space, it is inspired by the cities of Limerick and Cork, particularly the language of these cities.
2. Barry 2012, n.p.
3. Hamill 2012, n.p.
4. Barry 2012, n.p.
5. Bohane 2012, n.p.
6. Barry 2012, n.p.
7. Dryden 2003, 188–189.
8. Dryden 2003, 188.
9. Mieville 2009, 512.
10. Minter 2013, n.p.
11. Minter 2013, n.p.
12. Dryden 2003, 19.
13. Barry 2012, n.p.
14. Barry 2012, n.p.
15. Barry 2012, n.p.
16. Barry 2012, 3.
17. Barry 2012, 3.
18. Barry 2012, 3.
19. Barry 2012, 3.
20. Barry 2012, 3.
21. Barry 2012, 3.
22. Barry 2012, 3.
23. Barry 2012, 3.
24. Barry 2012, 3.
25. Barry 2012, 3.
26. Barry 2012, 3.
27. Barry 2012, 4.
28. Barry 2012, 4.
29. Barry 2012, 4.
30. Barry 2012, 4.
31. Barry 2012, 4.
32. Barry 2012, 4.
33. Barry 2012, 4.
34. Barry 2012, 5.
35. Barry 2012, 5–6.
36. Barry 2012, 6.
37. Barry 2012, 6.
38. Spooner 2006, 8.

39. Barry 2012, 4.
40. Dryden 2003, 19.
41. Barry 2012, 18.
42. Barry 2012, 18.
43. Barry 2012, 18.
44. Barry 2012, 7.
45. Barry 2012, 7.
46. Barry 2012, 7.
47. Barry 2012, 7.
48. Baldick 1992, xiii.
49. Bissett 2001, 5.
50. Barry 2012, 7.
51. Barry 2012, 7.
52. Barry 2012, 7.
53. Barry 2012, 10–11.
54. Barry 2012, 10.
55. Barry 2012, 10.
56. Barry 2012, 10–11.
57. Barry 2012, 11.
58. Barry 2012, 11.
59. Spooner 2006, 13.
60. Barry 2012, 20.
61. Barry 2012, 20.
62. Barry 2012, 20.
63. Barry 2012, 20.
64. Barry 2012, 20.
65. Platt 2001, 1.
66. Barry 2012, 69.
67. Barry 2012, 69.
68. Barry 2012, 69.
69. Barry 2012, 69.
70. Barry 2012, 73.
71. For a more detailed discussion of H.P. Lovecraft, China Mieville and the City, please see Colebrook, Martyn 'Comrades in Tentacles: H.P. Lovecraft and China Mieville' in New Critical Essays on H.P. Lovecraft (2013) (ed. David Simmonds). Palgrave Macmillan, US.
72. Barry 2012, 95.
73. Barry 2012, 95.
74. Barry 2012, 79.
75. Barry 2012, 79.
76. Barry 2012, 89.
77. Barry 2012, 116.
78. Barry 2012, 116.
79. Barry 2012, 116.

References

Baldick, Chris. *The Oxford Book of Gothic Tales.* Oxford: Oxford University Press, 1992 [2009].

Barry, Kevin. *City of Bohane.* London: Vintage, 2012a [2011].

———. 'Paperback Q&A.' *The Guardian*, March 27, 2012b. https://www.theguardian.com/books/2012/mar/27/paperbook-kevin-barry.

Bissett, Alan. 'The Dead Can Sing: Introduction.' In *Damage Land: New Scottish Gothic Fiction*, ed. Alan Bissett, 1–8. Edinburgh: Polygon, 2001.

Colebrook, Martyn. 'Comrades in Tentacles: H.P. Lovecraft and China Mieville.' In *New Critical Essays on H.P. Lovecraft*, ed. David Simmonds, 209–226. New York: Palgrave Macmillan, 2013.

Dryden, Linda. *The Modern Gothic and Literary Doubles: Stevenson, Wilde and Wells.* Basingstoke: Macmillan, 2003.

Duncan, Hal. *Vellum.* London: Macmillan, 2005.

Gray, Alasdair. *Lanark: A Life in Four Books.* Edinburgh: Canongate, 1981 [1985].

Hamill, Pete. 'Auld Times.' *The New York Times*, March 29, 2012. http://www.nytimes.com/2012/04/01/books/review/city-of-bohane-by-kevin-barry.html. Accessed April 1, 2012.

Lee, Jonathan. 'Jumping Off a Cliff: An Interview with Kevin Barry.' *The Paris Review*, November 12, 2013. Web. Accessed June 17, 2016.

Mieville, China. *Perdido Street Station.* London: Pan Books, 2002.

Miéville, China. 'Weird Fiction.' In *The Routledge Companion to Science Fiction*, ed. Mark Bould and Sherryl Vint, et al., 510–515. London: Routledge, 2009.

Platt, Len. *Aristocracies of Fiction: The Idea of Aristocracy in Late-Nineteenth-Century and Early-Twentieth-Century Literary Culture.* Westport: Greenwood Press, 2001.

Raban, Jonathan. *Soft City.* London: Hamish Hamilton, 1974.

Sage, Victor. 'Irish Gothic: C.R. Maturin and J.S. Le Fanu.' In *A New Companion to The Gothic*, ed. David Punter, 135–147. London: John Wiley & Sons, 2015.

Smethurst, Paul. *The Postmodern Chronotope: Reading Space and Time in Contemporary Fiction.* Atlanta, GA: Editions Rodopi B.V., 2000.

Spooner, Catherine. *Contemporary Gothic.* London: Reaktion Books, 2006.

CHAPTER 9

John Banville: The City as Illuminated Image

Neil Murphy

John Banville has always insisted that his fiction contains no social or political messages, doesn't directly engage with the material landscape of his personal experiences or social milieu, and has instead conjured resonant topographical literary metaphors. Nonetheless, all of the novels after *Kepler* (1981), except *The Untouchable* (1997) and *Shroud* (2002), are nominally located in Irish settings, with many set in urban environments ranging from Dublin to Rosslare, Co. Wexford, even if the precise identities of the locations are rarely overtly declared. Banville's treatment of these urban spaces mirrors his representation of reality, more generally, which is accessible only via the language-obsessed subjective consciousnesses that dominate his fictions. As he has indicated in an interview, 'The world is not real for me until it has been pushed through the mesh of language',[1] and the texture of that aestheticised world is, in turn, signified only as an artefact. This chapter will demonstrate the manner in which Banville's illuminated aesthetic images of his urban contexts project ghostly traces of their origins while always, nonetheless, declaring their invented otherness. As the philosopher Gordon Graham argues: '… a novel is not to be thought of as providing us with a faithful reflection of experience or a skillful summary of it, but as obliging us to view some aspect of experience through an image which allows us to attain an

N. Murphy (✉)
Nanyang Technological University, Singapore, Singapore
e-mail: camurphy@ntu.edu.sg

M. Beville, D. Flynn (eds.), *Irish Urban Fictions*, Literary Urban Studies, https://doi.org/10.1007/978-3-319-98322-6_9

illuminating perspective upon it'.[2] While Banville's aesthetic illuminations repeatedly point to the complex problem of representation, they also permit one to gain a resonant sense of the contexts from which they are derived and always seek to illuminate the merely material. In discussing the potent evocation of 'cityscapes in the industrial age',[3] David Lodge draws attention to 'the vivid visual detail of Dickens's London'[4] in order to demonstrate the manner in which his fiction adjusted to a post-Romantic age. However, Elana Gomel has also convincingly argued that there are two distinct diegetic levels at work in Dickens' engagement with London:

> On the one hand, there is the mimetic urban space that Dickens shared with his readers, evoked through familiar place names: Lincoln's Inn Hall, Holborn Hill, and Temple Bar. In this mimetic space, the polluted streets are covered with physical dirt, the unwholesome air is laden with actual fog, and the Court of Chancery is a solid building, housing corrupt bureaucrats.[5]

This is akin to Lodge's Dickens, but Gomel's second narrative level is perhaps more crucial. 'The fantastic extradiegetic space', she argues, is 'self-consistent but topologically incompatible with the primary diegetic level…', although it still 'impacts the lives and actions of the characters inhabiting the realistic diegetic space'.[6] Dickens' fantastic co-exists with the realist writer with whom most Victorian critics have been engaged, in part because he so effectively maintained the mutually beneficial narrative doubling in his major novels, with the balance in favour of the primary diegetic level of realist representation always foregrounded. That balance, in Banville, as I will indicate, is a rather more complex affair, with the primary narrative level being far less committed to even the semblance of an accurate depiction of the cities and towns in which they are nominally located. Gomel suggests that the problem of representation in postmodernity, in general, is that the story of the city 'becomes unreadable, haunted by the memory of the cataclysm and the premonition of its return. And so narrative struggles with the impossible city'.[7] But her sense is of immense anxiety and of some form of aesthetic and epistemological crisis. This, in many respects, speaks to the anxiety of representation that one finds throughout modernist fiction, everywhere from Joyce's protean Sandymount beach in *Ulysses*, to *Mrs. Dalloway*'s London, or Eliot's London wastelands. Many postmodern texts, in fact, do not exhibit this kind of anxiety at the vanishing of certainty in knowing or representing unknown landscapes. The nouveau-roman novelist, Alain Robbe-Grillet, for example, insists on the disappearance of certainty as a liberation: 'And the man of today (or of tomorrow...) feels no sense of deprivation or affliction at this absence of meaning. He no longer feels lost at the idea of such a vacuum. His heart no longer needs to take refuge in an abyss'.[8] In fact, postmodern novels like Calvino's *Invisible Cities* and *If on a Winter's Night a Traveler*, as well as much of Nabokov, Flann O'Brien, Brautigan, and Barth, offer up deeply inventive and playful urban landscapes throughout their work, rather than anxiety-driven narratives of lost moorings.

John Banville reveals much of that playful spirit in his early work, with *Nightspawn* (1971) and *Birchwood* (1973) featuring shimmering houses, an exploding granny, parodic treatments of crime fiction, a fantastic version of the big house novel, and a subversion of all attempts to designate a 'reality' derived from a verifiable encyclopaedic reality. But Banville's later work offered up an immensely varied series of approaches to the question of context, from the detailed historical reconstructions of sixteenth- and seventeenth-century Wloclawek, Cracow, and Rome (*Doctor Copernicus,* 1976) and Graz, Prague, and Linz (*Kepler*) to the return to a more sedately imagined big house in *The Newton Letter* (1981). Banville's construction of fictional storyworlds thereafter has become increasingly complex in terms of its layering of multiple diegetic levels but the texture of the primary narrative levels, on which a nominal engagement with context is evident, is perhaps where a fundamental difference between his work and that of realist novelists like Dickens, and that of more anxiety-dominated modernist writers, is clear.

Many critics have pointed to the recurring pattern of the creative challenge offered by postmodern city-spaces. Stefan. L. Brandt, for example, citing Soja, suggests that the postmodern metropolis 'is spatially organised through strategies of fragmentation, and decentering'[9] and that postmodern texts that seek to engage with such spaces frequently 'point to the liminality of the city experience—that is, to the transitory and ultimately invasive character of postmodern urbanity'.[10] James Donald, similarly, claims that there 'is no such thing as a city' and suggests that it instead emerges as a space redolent with the 'interactions of historically and geographically specific institutions, social relations production and reproduction, practices of government, forms and media of communication and so forth. The city, then, is above all a representation. I would argue that the city constitutes "an imagined environment"'.[11] Maria Beville, in turn, considers the postmodern city a 'scene of uncanny spectrality, an uncanny media-landscape to be found in many contemporary texts'.[12] There is little doubt as to the accuracy of such observations, particularly with respect to authors like Auster, DeLillo, Baricco, and other contemporary writers who directly engage with the complex imaginaries that are postmodern cities.

Nonetheless, Banville's urban contexts in his post-2000s fiction present significantly different propositions, even as they ultimately resist the representative impulse that one discerns even in Auster's postmodern New York. Rather than map bewitching media-scapes or the architectural confusion of contemporary metropolises, Banville instead positions the subjective imagining mind at the centre of all representative efforts. Timothy Yu's consideration of the manner in which Leopold Bloom responds to the streets on Dublin is partially indicative of the response that one witnesses in many of Banville's narrator-heroes. Yu details the obvious to some degree when he suggests that the identification of street names and products gradually gives way to Bloom's reverie of the 'east' and points to the hero's awareness of the precise source of his reverie: Frederick Thompson's *In the Track of the Sun* (1893). He thus registers an awareness that there are different layers of the real at play; a sense of the multiplicity of levels of fictional reality is powerfully conveyed.[13] This is also a feature of all of Banville's mature fiction, although, as we shall see, the clear demarcation between these levels is not as starkly obvious as it is in Bloom's alternating between an authentic reality and his 'imagined' reveries. It is very clear throughout all of Banville's work, across different genres, that no such clean distinction is feasible.

Banville has occasionally acknowledged that there is 'something of me and my history' in all of the novels,[14] but the relationship between his fiction and life is muted, at best, with his basic biography offering little direct correlation with the subject matter of his novels. Born in Wexford, he attended St Peter's College in the same town and later worked for the Irish air carrier *Aer Lingus* for several years before working as a journalist for close to 40 years with the now-defunct *The Irish Press* and *Irish Times*. While he acknowledges that *The Sea* (2005) is located in a fictionalised Rosslare, the seaside town where he spent every summer as a child,[15] and Cleave's house in *Eclipse* is the house in which he was born,[16] he insists that he has no interest in writing about himself, primarily because he views art to be an impersonal pursuit.[17] In *Time Pieces: A Dublin Memoir*, Banville comes as close as he ever has to writing about his own life and acknowledges, in particular, that certain places, locations, 'imprint themselves on the memory with improbable vividness and clarity'.[18] He reveals how a certain bend in the Avoca river found its way into *The Newton Letter*[19] and a flat in Mount Street in Dublin, where he once lived, is Quirke's dwelling place in the Benjamin Black novels. Nevertheless, Banville's memoir also insists on the essential elusiveness of the world about which he writes: 'Let us say, the present is where we live, while the past is where we dream'.[20] More emphatically, he insists in the memoir that he has never 'paid much attention to my surroundings wherever it was I happened to find myself. ... Art is a constant effort to strike past the daily doings of humankind in order to arrive at, or at least to approach as closely as possible to, the essence of what it is, simply, to be'.[21] Speaking of Dublin, where he has lived almost continuously since his late teens, Banville concedes that he did not have much interest in the city, nor did he feel that it was even his to possess in any form:

> The truth is, I had little interest in Dublin's past, and not much in its present either, if it came to that. ... For me, as a writer in the making, the fact was that Joyce had seized upon the city for his own literary purposes and in doing so had used it up, as surely as Kafka did with the letter K, and consequently the place was of no use to me as a backdrop for my fiction. True, some of my early short stories take place in an identifiable Dublin, but they could just as well have been set in London, or Paris—or Moscow for that matter. It was not until much later, when I invented my dark brother Benjamin Black, that I saw the potential of 1950s Dublin as a setting for his *noir* novels.[22]

The Benjamin Black observation aside, to which we will presently return, Banville's apparent disinterest in Dublin as a possible landscape for his fictions is more significant as an aesthetic position than it is a refusal to work in the shadow of Joyce. Inherent in his decision is something of the ambivalence that one increasingly finds in Banville, between the insistence on the impossibility of 'true' representation and the doubtful hesitation that so frequently accompanies this insistence: 'I seek to comfort and perhaps even exonerate myself with the thought that this is what artists do, the imagination being the only true place in which richly to live. But am I convinced?'.[23] The hesitation speaks of his perpetual desire to somehow, perhaps paradoxically, attend to the notion of the real, in the face of his disbelief in its representation. The key appears to be located in his sense of an illuminated imaginative process. A similar reaction is evident in his travel book, *Prague Pictures* (2003), particularly when he considers the quality of his representation of Prague in his early novel, *Kepler*, which was written before he'd ever visited the Czech capital city:

> Yes, I had got it right, to a startling degree. Why was I not pleased? In part because, standing there surveying my handiwork, I was struck again by the essential fraudulence of fiction. Conjure a winter morning, a river and a castle and a traveller disembarking with a book under his arm, and for the space of a page or two an implied world comes to creaky life. It is all sleight of the imagination, a vast synecdoche. And yet one goes on doing it, spinning yarns, trying to emulate blind Fate herself. (9)

If the fiction is fraudulent, it follows that the world depicted in there is also somehow fraudulent, and this, paradoxically, is the key to its aesthetic achievement. Tim Adams, in his review of Banville's travel book, suggests that 'It is entirely appropriate that John Banville described Prague before he ever went there. All cities, to some extent, live mostly in our heads, but the Czech capital often seems entirely an imaginative construct'.[24] Banville admits as much throughout *Prague Pictures*, offering up something of an extended love letter to a city that he imagines and reimagines, in the process offering it renewed life, via the intensity and quality of that imagining.

Banville's attitude to being considered an Irish writer is closely related to his response to such world-invention tendencies: 'I think it's not that writers should be writing about Ireland and our crisis and so on. That's not what artists do. I don't think it is. I don't think it's even what novelists do'.[25] Furthermore, he establishes a clear distinction between the values held by the man/woman and the writer:

> Away from his desk the novelist can care deeply about the social, political, moral aspects of what he is writing but when he sits down to write, all those concerns fall away and nothing matters except the putting down of one carefully chosen word after another carefully chosen word, until a sentence is finished, then a paragraph, then a page, then a chapter, then a book.[26]

The recurring observation that subject matter, or content, is only of peripheral significance is, of course, closely related to Banville's refusal to write fiction that is explicitly derived from his life or that of his general social milieu: 'By what means, then, does fiction *get at* the world? Not by engagement, I am convinced, but precisely by disengagement, by adopting a posture of bland innocence, standing back with empty palms on show… The subject matter hardly matters'.[27]

And yet Banville does write about the world, specifically about Dublin, in his Benjamin Black novels,[28] which represent the only sustained writing about the city in Banville's fiction. He acknowledges that the 'city is God's gift to the crime writer' and that 'the urban wilderness lends itself with particular aptness to *noir* fiction, whether it be Maigret's Paris, Philip Marlowe's Bay City, a lightly fictionalised version of Santa Monica, or Dostoevsky's St Petersburg'.[29] Andrew Kincaid has also written about the intimate connection between major American cities and the novels of Hammel and Chandler, particularly in terms of their 'shared mood and atmosphere: dark, downbeat, paranoid, and pessimistic'.[30] In Ireland, in a similar vein, Kincaid discerns the emergence of a potent socially inclined variation of the great American *noir* fictions: '… over the last decade an interesting new version of noir, perhaps a hybrid of hardboiled detective, forensic thriller and crime mystery, has arisen in Ireland, a form that reflects the pace of change while also giving nervous expression to the disillusionment and self-doubt that lay just below the glittering prosperity'.[31] While a distinct form of the crime novel may well have emerged in Ireland, in response to the particularity of the Celtic Tiger context, and its aftermath, at least some of these generic aspects can be found in crime fiction more generally. For example, Jon Thompson argues that the crime genre offers

> myths of the experience of modernity, of what it is like to live in a world dominated by the contradictory forces of renewal and disintegration, progress and destruction, possibility and impossibility. The capacity of crime fiction to evaluate different historical moments in the experience of modernity is not an accidental feature; rather, it is a dominant convention of the genre.[32]

While Benjamin Black's novels offer a challenging test case for Thompson's and Kincaid's observations, given Banville's attitude to the non-social role of art, the central significance of the city in Black's work remains. The modern urban city is a fundamental part of the aesthetic veneer of the hard-boiled detective novel and, as John Scaggs observes, citing Willet, the 'modern city of hard-boiled fiction is 'a wasteland devastated by drugs, violence, pollution, garbage and a decaying physical infrastructure,' and it is down the mean streets of this urban wasteland that the private eye must go…'.[33] Nonetheless, writers like Chandler, and Benjamin Black in the Quirke novels, operate in fictional landscapes in which the ontological conditions of their heroes are continually foregrounded. The brooding sense that justice is never ultimately achieved, irrespective of plot resolution, permeates the experiences of Quirke just as it does Chandler's Marlowe. In fact, Scaggs' summary of the unintelligible world of the hard-boiled detective novel posits that the achieved order in the crime plot co-exists with a broader sense of disconnectedness:

> The fakery and artifice that characterise the modern city of hard-boiled fiction drive a wedge between what is seen and what is known, and in this way the private eye's quest to restore order becomes a quest to make sense of a fragmented, disjointed, and largely unintelligible world by understanding its connections, or, more often, its lack of connections.[34]

Such an argument implies that engagement with a contemporary urban context serves an alternate function to the relatively prescriptive elements of the genre. Fintan O'Toole has similarly argued that Irish crime fiction, in general, in recent years 'has become arguably the nearest thing we have to a realist literature adequate to capturing the nature of contemporary society'.[35] Whether Black's novels fit O'Toole's realist model is certainly a contentious point. For example, when Hugh Haughton asks if he is correct in detecting 'a more politically conscious writer about the historical Ireland'[36] in the Benjamin Black novels, Banville forcefully disagrees:

> Well, I wish that I could agree with you, but I don't feel that. I really believe that any writer who imagines he has a social voice is in trouble. If you mix politics and art, you get bad politics and bad art. I can imagine, of course, if, for instance, it seemed to me tomorrow that the civil war in the North would make a subject for a novel for me, that I could write it as a novel rather than a commentary on war, then I would do it, but that would be incidental. I think this is always true. For art, subject is always incidental, or at least secondary, to the work itself.[37]

These observations extend the familiar position that Banville has occupied throughout his career, insisting that subject matter, or ethical positioning, is secondary to formal significance. In the Banville novels, in general, this position can be validated—despite alternative perspectives from some critics.[38] But Haughton's observations with respect to the Quirke novels merit some consideration. The brooding presence of Dublin is everywhere felt in the Black novels, as in *Elegy for April* (2010), which opens with a detailed urban description that is simultaneously nightmarish and unmistakably identifiable as the Irish capital:

> For days a February fog had been down and showed no sign of lifting. In the muffled silence the city seemed bewildered, like a man whose sight has suddenly failed. People vague as invalids groped their way through the murk, keeping close to the house-fronts and the railings and stopping tentatively at street corners to feel with a wary foot for the pavement's edge. Motor cars with their headlights on loomed like giant insects, trailing milky dribbles of exhaust smoke from their rear ends. The evening paper listed each day's toll of mishaps. There had been a serious collision at the canal end of the Rathgar Road involving three cars and an Army motorcyclist.[39]

Or, more evocatively, if perhaps less noir, *Even the Dead* (2015) details the mood of the city streets in clearly realist terms:

> Girls in Summer dresses were walking by the river, and there were swans on the water, and flags were rippling in the warm breeze, and in Grafton street there would be the rich brown smell of roasting coffee beans from the open doorway of Bewley's Oriental Café, and paper boys would be calling out the latest headlines, and there would be the sound of horses' hoofs on cobblestones, and the cries of the flower-sellers at their stalls. Summer. Crowds. Life.[40]

While this hardly amounts to atomistic description, it does offer realist engagement with Dublin. Furthermore, in this ontological context, throughout all the Quirke novels a general air of corruption circles around the Catholic church. Many of the crimes with which Quirke becomes embroiled are played out in the urban shadows of 1950s Dublin and are connected in some way to the dark presence of the church and its role in the corridors of power.

In fact, a sense of moral objection is also occasionally evident in Banville's interviews, in which he likens the enslavement of the Irish Catholic church in the 1950s, and beyond, to the fate of countries behind the Iron Curtain:

> Of course, we were under the yoke of an iron ideology. We were told in those days that the Soviet satellite countries, behind the Iron Curtain, they are not free there. We are free. It was only in the '90s when we actually did free ourselves that we realized for all those years we were exactly like a Soviet country. The church and state were hand in hand just as the state and party were hand in hand there. And our lives were completely un-free.[41]

Near-identical observations also find expression in the Black novels with Quirke at one point warning Phoebe about the nature of the Catholic church: 'They're just the same as the Communists they're always warning us about—two sides of the same coin'.[42] And yet, Banville rejects the suggestion that the Quirke novels are socially motivated and argues that the church-linked corruption is 'just material' and denies, with respect to *Christine Falls*, that he had 'a crusading social purpose', explaining that he 'just wanted to write a novel, and the scandals that had just begun to be revealed at that time seemed ideal for my purpose, as they continued to be'.[43] The function of the setting is thus a function of genre, rather than of ethics. It also explains his motivation in offering a detailed context linked to Church-sponsored adoptions and orphanages: 'I wanted to give him a past that was shadowed. First of all, to make him an orphan so he wouldn't know his background because that's the source of his curiosity, his urge to know, to find out secrets', but he also admits that a new awareness in Ireland of the horrors perpetrated by the Catholic church 'was preying on my mind'[44] (cited in Ruland), and concedes that 'we're never free of our time. We like to think we are but we're not'.[45]

Whatever the significance of Banville's conscious or unconscious levels of ethical awareness and reaction, the surface representative texture of Black's Dublin frequently gives way to a more nuanced and unstable context.[46] One is repeatedly reminded of the co-existence of different layerings of reality, as in when Quirke visits Bewley's Cafe, only to discover the extraordinary foreignness of literal reality: 'It seemed so long ago, that time, a kind of sun-dappled antiquity, as if it were an Attic glad he was remembering and not a shabby and overcrowded café in a faded little city with a past that felt more immediate that its present'.[47] The instability of material reality that is so frequently a feature of Banville's fiction is clearly evident in Black's work. For example, returning to his flat in *Elegy for April* after spending time in St John's drying-out centre for recovering alcoholics, Quirke marvels at the 'suddenly estranged and sullen surroundings',[48] and more recently, in *Even the Dead*, during Phoebe's meditation on the essential difference between objects and humans:

> It occurred to her how dissimilar things were from people [...] but objects were always obstinately themselves. Or, no, not 'obstinately', that was the wrong word. 'Indifferently', that was what she meant. She recalled what her father had said to her once, long ago, in the days before she knew he was her father. 'The thing to remember,' Phoebe,' he had said, 'is that the world is indifferent to us and what we do.' ... she had never forgotten him saying it.[49]

The world will always remain other, and while Banville's first-person protagonists generally reflect more self-consciously on their contexts than Black's characters, their intuitive awareness of the unyielding reality is the same.

At key moments like these, a sense of the instability of ontological reality pervades Banville's fictional worlds, just as it does those of Black. In the novel *Eclipse*, for example, Alex Cleave at one point ponders the nature of the relationship between material reality and the images we construct in response to that reality:

> If the [fictional] ghostly scene has a chair in it, say, that the woman is sitting on, and that occupies the same space as a real chair in the real kitchen, and is superimposed on it, however ill the fit, the result will be that when the scene vanishes the real chair will retain a sort of aura, will blush, almost, in the surprise of being singled out and fixed upon, of being lighted upon, in this fashion.[50]

The result is a process of doubling that, in Banville, projects a highly aestheticised surface that bears traces of the materiality from which it is derived and in turn leaves an afterglow, a quality of aesthetic strangeness, in its wake: 'Everything is strange now. The most humdrum phenomena fill me with slow astonishment'.[51] Dublin itself is subjected to this self-consciously transformative quality at times, in particular in the haunting opening scene of the novel *Athena* (1995):

> Strange how the city becomes deserted at this evening hour. … A car creeps up on me from behind, tyres squeaking against the sides of the narrow footpaths, and I have to stop and press myself into a doorway to let it pass. … It shoulders by me with what seems a low chuckle and noses down an alleyway, oozing a lazy burble of exhaust smoke from its rear end, its lollipop-pink tail-lights swimming in the deliquescent gloom.[52]

The world, again suffused with strangeness, contains traces of itself while simultaneously asserting itself as image in the fictionalised storyworld. This tension between the real and the images that are conjured from its essence becomes increasingly apparent in Banville's work after 2000. In *The Sea* (2005), for example, the central tension of the novel emerges precisely from this contrast:

> I found that the model of the house in my head, try as it would to accommodate itself to the original, kept coming up against a stubborn resistance. Everything was slightly out of scale, all angles slightly out of true [...] I experienced a sense almost of panic as the real, the crassly complacent real, took hold of the things I thought I remembered and shook them into its own shape. Something precious was dissolving and pouring away between my fingers. Yes how easily, in the end, I let it go. The past, I mean the real past, matters less than we pretend.[53]

The distinction between the image and the world from which it is derived is everywhere apparent in Banville's mature work, in fact the co-existence of the traces of both ensure that one is forever conscious of a certain doubling of intent. In turn, the quality of that doubling is the source of much self-reflexive fascination and perpetual expression of the otherness of the imagined variations.

In her discussion of the arts, the aesthetic philosopher Susanne Langer claims that the quality of otherness is central to art: 'Every real work of art has a tendency to appear thus dissociated from its mundane environment. The most immediate impression it creates is one of "otherness" from reality—the impression of an illusion enfolding the thing, action, statement, or flow of sound that constitutes the work'. This otherness, she argues, has been variously described as 'strangeness', 'semblance', 'illusion', 'transparency', 'autonomy', or 'self-sufficiency'.[54] Banville too has frequently voiced his conviction that the world's strangeness, and his characters' puzzlement in the face of it, best defines his work,[55] and that strangeness is the mark of art. Langer's sense of the image is as a replacement presence rather than a representation. The image, unlike the world from which it is derived, contains the qualities of strangeness and otherness that identify the work of art as different: 'An image in this sense, something that exists only for perception, abstracted from the physical and causal order, is the artist's creation'.[56] Its appearance is abstracted from its material existence, should it have a counterpoint in reality (not all works of art have direct counterparts in the real). Banville's and Black's urban centres and spectral worlds repeatedly convey a sense of their own otherness, through the simultaneous shimmering presence of Dublin, Wexford, and so on in the storyworlds. The towns and cities that nominally emerge in his fictions, and arguably in his memoir and travel writing, exist primarily as illuminated replacement images of a material world experienced but never explicitly known.

Notes

1. Belinda McKeon, *The Paris Review*. 'John Banville, The Art of Fiction No. 200.' Accessed April 24, 2018.
2. Gordon Graham, *Philosophy of the Arts: An Introduction to Aesthetics*. New York: Routledge, 2007, p. 144.
3. David Lodge, *The Art of Fiction*. Harmondsworth: Penguin 1992, p. 59.
4. Ibid., p. 57.
5. Elana Gomel, *Narrative Space and Time: Representing Impossible Topologies in Literature*. New York: Routledge, 2014, p. 41.
6. Ibid., p. 41.
7. Ibid., p. 173.
8. Alain Robbe-Grillet, 'Snapshots and Toward a New Novel.' *Revolutions of the Word: Intellectual Contexts for the Study of Modern Literature*. Ed. Patricia Waugh, London: Arnold, 1997, pp. 292–293.
9. Stefan L. Brandt, "The City as Liminal Space: Urban Visuality and Aesthetic Experience in Postmodern U.S. Literature and Cinema," in *Amerikastudien/American Studies*, Vol. 54, No. 4 (2009), p. 553.
10. Ibid., p. 555.
11. James Donald, 'Metropolis: The City as Text.' *Social and Cultural Forms of Modernity*. Ed. Robert Bocock and Kenneth Thompson. Cambridge, UK: Open University, 1992, p. 422.
12. Maria Beville, 'Zones of Uncanny Spectrality: The City in Postmodern Literature. *English Studies*, 94:5 (2013): p. 604.
13. Timothy Yu, 'Oriental cities, postmodern futures: Naked Lunch, Blade Runner, and Neuromancer.' *Melus* 33:4 (Winter 2008), pp. 45–71.
14. 'Fully Booked: Q & A with John Banville.' Interview by Travis Elborough. *Picador*, June 29, 2012. https://www.panmacmillan.com/blogs/picador/fully-booked-q-a-with-john-banville.
15. Belinda McKeon, 'John Banville: The Art of Fiction No. 200.' *The Paris Review* No. 188 (Spring 2009) (accessed October 31, 2016).
16. 'Oblique Dreamer: Interview with John Banville.' *The Observer* (17 September 2000), https://www.theguardian.com/books/2000/sep/17/fiction.johnbanville (accessed October 31, 2016).
17. Elborough, 'Fully Booked'.
18. John Banville, *Time Pieces: A Dublin Memoir* 2016, p. 2.
19. John Banville, The Newton Letter. 1982, p. 2.
20. John Banville, *Time Pieces: A Dublin Memoir* 2016, p. 4.
21. Ibid., p. 53.
22. Ibid., p. 81.
23. Ibid.

24. John Banville, 'Reality Czech: John Banville goes right to the heart of inner-city living in his impressionistic hymn to the Czech capital, Prague Pictures. Interview with Tim Adams, Sunday 21 September 2003. https://www.theguardian.com/books/2003/sep/21/travel.travelbooks.
25. Haughton and Radley, 'Interview with John Banville', p. 867.
26. McKeon, 'John Banville: The Art of Fiction…'.
27. Schwall, p. 119.
28. Immediately after winning the Booker Prize for his novel, *The Sea*, John Banville surprised the literary world by publishing his first 'Quirke' crime novel, *Christine Falls* (2006), under the pseudonym Benjamin Black. He has since published a further nine crime novels, six of which retain the central figure, Quirke, the chief pathologist in the Holy Family hospital in Dublin of the 1950s, a profession that inadvertently allows him access to the crime world.
29. 'Take a trip to Sin City: John Banville on why the best crime fiction is urban', *The Telegraph*, 4 June 2017. http://www.telegraph.co.uk/books/what-to-read/take-trip-sin-city-john-banville-best-crime-fiction-urban/.
30. Andrew, Kincaid, "'Down These Mean Streets': *The City and Critique in Contemporary Irish Noir*," *Éire-Ireland*, Volume 45:1&2 (Earrach/Samhradh/Spring/Summer 2010): p. 40.
31. Ibid., p. 41.
32. Jon Thompson, *Fiction, Crime, and Empire: Clues to Modernity and Postmodernity*. Urbana and Chicago: University of Illinois Press, 1993, p. 2.
33. John Scaggs, *Crime Fiction*. London: Routledge, 2005, p. 70.
34. Ibid., p. 72.
35. Fintan O'Toole, 'From Chandler and the 'Playboy' to the contemporary crime wave,' *Irish Times* (Saturday, November 21, 2009). http://www.irishtimes.com/culture/tv-radio-web/from-chandler-and-the-playboy-to-the-contemporary-crime-wave-1.776393 (accessed August 26, 2017).
36. 'An Interview with John Banville,' Interview by Hugh Haughton and Bryan Radley. *Modernism/Modernity* 18, no. 4 (November 2011), p. 867.
37. Ibid., p. 868.
38. For example, Derek Hand insists on the essential Irishness of Banville's work (Hand, 2009, pp. 12–19), and John Kenny has considered the merits of historicising Banville's work (Kenny, 2009, pp. 55–57).
39. John Banville. *Elegy for April*. 2010. p. 3.
40. John Banville. *Even the Dead*. 2015. p. 1.
41. 'Guest Profile: Benjamin Black Parts 1–4.' Interview with Jim Ruland. *The Elegant Variation: A Literary Weblog*, December 8, 2008.
42. John Banville. *Even the Dead*. 2015. p. 194.

43. 'Q & A with Benjamin Black,' *Crimespree Magazine*, January 27, 2016. http://crimespreemag.com/qa-with-benjamin-black/ (accessed August 25, 2017).
44. 'Guest Profile: Benjamin Black Parts 1–4.' Interview with Jim Ruland. *The Elegant Variation: A Literary Weblog*, December 8, 2008.
45. 'Oblique Dreamer.' Interview with *The Observer*. *The Guardian* (September 17, 2000). https://www.theguardian.com/books/2000/sep/17/fiction.johnbanville (accessed August 25, 2017).
46. For a more comprehensive treatment of Benjamin Black's treatment of fictional reality, please refer to my article 'Crimes of Elegance: Benjamin Black's Impersonation of John Banville,' in *Moving Worlds* (Leeds). Spring 2013: pp. 19–32.
47. John Banville. *A Death in Summer.* 2011. p. 216.
48. John Banville. *Elegy for April.* 2010. p. 34.
49. John Banville. *Even the Dead.* 2015. p. 179.
50. John Banville, Eclipse. 1995. p. 48.
51. Ibid., p. 52.
52. John Banville. *Athena.* 1995. p. 1.
53. John Banville. *The Sea.* 2005. pp. 156–157.
54. Susanne Langer, *Feeling and Form* (New York: Charles Scribner's Sons, 1953), pp. 45–46.
55. Hedda Friberg, 'John Banville and Derek Hand in Conversation,' *Irish University Review* 36, no. 1 (2006): p. 206.
56. Langer, p. 47.

References

Banville, John. *Birchwood*. London: Panther Books, 1973.

———. *Kepler.* London: Panther Books, 1985.

———. *Doctor Copernicus.* London: Secker & Warburg, 1976. London: Minerva, 1990.

———. *The Newton Letter.* An Interlude. London: Seeker & Warburg, 1982. London: Minerva, 1992.

———. *Nightspawn*. London: Seeker & Warburg, 1971. Loughcrew: Gallery Press, 1993.

———. *Athena*. London: Seeker & Warburg, 1995a.

———. *Eclipse*. London: Secker & Warburg, 1995b.

———. *Shroud*. London: Picador, 2002a.

———. *The Untouchable*. London: Picador, 2002b.

———. *Prague Pictures: Portraits of a City.* London: Bloomsbury, 2003a.

———. 'Reality Czech: John Banville Goes Right to the Heart of Inner-City Living in His Impressionistic Hymn to the Czech Capital, Prague Pictures.'

Interview with Tim Adams Sunday, September 21, 2003b. https://www.theguardian.com/books/2003/sep/21/travel.travelbooks.
———. *The Sea*. London: Picador, 2005.
———. 'An Interview with John Banville.' Interview by Hugh Haughton and Bryan, 2011a.
———. Radley. *Modernism/Modernity* 18, no. 4 (November 2011b): 855–869.
———. *A Death in Summer*. London: Mantle, 2011c.
———. *Time Pieces: A Dublin Memoir*. Dublin: Hachette Books, 2016.
———. 'Take a Trip to Sin City: John Banville on Why the Best Crime Fiction is Urban.' *The Telegraph*, June 4, 2017. http://www.telegraph.co.uk/books/what-to-read/take-trip-sin-city-john-banville-best-crime-fiction-urban/.
Beville, Maria. 'Zones of Uncanny Spectrality: The City in Postmodern Literature.' *English Studies*, 94, no. 5 (2013): 604.
Brandt, Stefan L. 'The City as Liminal Space: Urban Visuality and Aesthetic Experience in Postmodern U.S. Literature and Cinema.' *Amerikastudien/American Studies*, 54, no. 4 (2009): 553–581.
Donald, James. 'Metropolis: The City as Text.' In *Social and Cultural Forms of Modernity*, ed. Robert Bocock and Kenneth Thompson, 417–461. Cambridge: Open University, 1992.
Gomel, Elena. *Narrative Space and Time: Representing Impossible Topologies in Literature*. New York: Routledge, 2014.
Graham, Gordon. *Philosophy of the Arts: An Introduction to Aesthetics*. New York: Routledge, 2007.
Kincaid, Andrew. '"Down These Mean Streets": The City and Critique in Contemporary Irish Noir.' *Éire-Ireland*, 45, no. 1&2 (Earrach/Samhradh/Spring/Summer 2010): 39–55.
Langer, Susanne K. *Feeling and Form: A Theory of Art*. New York: Charles Scribner's Sons, 1953.
Lodge, David. *The Art of Fiction*. Harmondsworth: Penguin, 1992.
McKeon, Belinda. 'John Banville, The Art of Fiction No. 200.' *The Paris Review*, 2009. https://www.parisreview.org/interviews/5907/john-banville-the-art-of-fiction-no-200-john-banville. Accessed September 8, 2017.
Robbe-Grillet, Alain. 'Snapshots and Toward a New Novel.' In *Revolutions of the Word: Intellectual Contexts for the Study of Modern Literature*, ed. Patricia Waugh, 292–293. London: Ed. Arnold, 1997.
Scaggs, John. *Crime Fiction*. London: Routledge, 2005.
Thompson, Jon. *Fiction, Crime, and Empire: Clues to Modernity and Postmodernity*. Urbana and Chicago: University of Illinois Press, 1993.
Yu, Timothy. 'Oriental Cities, Postmodern Futures: Naked Lunch, Blade Runner, and Neuromancer.' *Melus* 33, no. 4 (Winter 2008): 45–71.

PART III

Cities of Change: Re-writing the City

CHAPTER 10

The Haunted Dublin of *Ulysses*: Two Modes of Time in the Second City of the Empire

Nikhil Gupta

'If we were all suddenly somebody else,' thinks Leopold Bloom in the 'Hades' episode of James Joyce's *Ulysses*.[1] More than a mere daydream, Bloom's reflection on metempsychosis comes while participating in the funeral rites of Paddy Dignam, and so the possibility of embodying otherness is wrapped up in the novel's preoccupation with a past that has not yet been put to rest in the colonial city featured in all of Joyce's fiction. In *Ulysses*, and in the 'Wandering Rocks' episode specifically, Joyce teases out these connections between overlapping times and overlapping experiences when he depicts the events occurring in Dublin on 16 June 1904 as unfolding according to the logic of two different temporal orders. On the one hand, coincidence structures the narrative to position individual characters alongside others and to render their shared moments significant in a style typical of national imagining. On the other hand, figural time marks Joyce's Dublin as a colonial city rooted in older forms of social order and structural power than the dictates of modern capitalism and national belonging that epitomise citizenship in the capitals of independent states. These two modes of time collide when the ghost of Charles Stewart Parnell returns to Dublin in 'Wandering Rocks.' When Parnell haunts Joyce's Dublin, he casts the city as a space decidedly out of sync with itself. Because Parnell's ghost appears through a figural order of time in *Ulysses*,

N. Gupta (✉)
University of Michigan, Ann Arbor, MI, USA

M. Beville, D. Flynn (eds.), *Irish Urban Fictions*, Literary Urban Studies, https://doi.org/10.1007/978-3-319-98322-6_10

his spectre evokes a sense of expectation and points to some other time, offering the colonial city a model for understanding its own sense of otherness and experience of political displacement.

With textual interpolations disrupting and connecting the 'Wandering Rocks' episode's 19 sections, the events noted in Dublin from three to four o'clock are connected to one another and synchronised according to the clock that runs throughout the novel's day-long duration. This version of simultaneity depicts, in Benedict Anderson's words, 'the idea of the nation, [...] a solid community moving steadily down (or up) history.'[2] In order to make sense of the interpolations that Joyce uses as a formal device to clock the movements of his Dubliners in 'Wandering Rocks,' the reader must move back and forth throughout the text. Intrusive lines early on are given an afterlife in later sections; only then are their fuller contexts revealed and, more often than not, an ironic link between the interrupted and intrusive scenes made clear. Deferral and recapitulation are the keys to mapping the city in this chapter of the novel; readers must allow themselves to get lost in sentences that jump without warning from place to place across Dublin, only later recognising how they connect. In 'Wandering Rocks,' then, Joyce renders Dublin as bound up in a mode of time representative of the nation; however, through the emergence of the shade of Parnell, Joyce also depicts the city as haunted by a political ghost whose return, according to the attendees at Paddy Dignam's funeral, had always already been prefigured. James Hansen has argued that the ghost of Parnell haunts the 'Scylla and Charybdis' episode of *Ulysses*, where Stephen Dedalus's lecture on *Hamlet* deploys memory as a form of historical revision, but locating the Irish politician's ghost in the 'Wandering Rocks' episode underscores the dual nature of community in Joyce's Dublin.[3] At one and the same time, the Irish of this novel connect to one another through the imaginative work that characterises national belonging, and they relate to one another through a type of city life that has developed according to the logic of colonialism. As Joyce's Dubliners go about their day together, their inner lives follow a figural pattern structured by memory and expectation. When Parnell's ghost becomes the focus of this temporal order marked by insistent recalling and constant waiting, the novel suggests a form of political longing in the colonial city that parallels the forms of sexual and commercial desire motivating and disappointing so many of the characters in Joyce's text. By setting Dublin, the second city of the British Empire, within two different modes of time in *Ulysses*, Joyce presents the view of Dublin that he has at the time of the novel's publication—a national capital that has yet to give up the ghost of its colonial oppression.

The textual interruptions that temporally link the 19 sections of 'Wandering Rocks' are a more pronounced example of the style that Joyce uses generally in his novel. Hugh Kenner shows how the world of *Ulysses* develops through delayed meaning. Different versions of the same event, he explains, expand the significance of that event for the reader and render the world of the text more substantial:

> Changing our earlier understanding with late facts, it resembles a plot like that of *Oedipus Rex*, where a terminal revelation alters all. But mutating each time it is reread, altering the very sense of early sentences as the import of later ones chances to come home, it resembles chiefly itself.[4]

Building on Kenner's reading of *Ulysses* as an experiment in the aesthetic of delay, Marjorie Howes argues that Joyce's novel not only anticipates moments coming later in the text, but it also looks back to the material it has presented earlier: 'The aesthetic of delay prompts the reader to revisit passages and solve riddles in the light of later information, while the aesthetic of parallax creates textual moments that are only intelligible in relation to what has been said before.'[5] Both Kenner and Howes emphasise Joyce's use of multiplicity and reiteration in bringing about the distinct narrative structure of his novel. *Ulysses* utilises this complex system of meaning because Joyce recognises both a synchronisable clock and a figural system of time at work in Dublin, the latter predicated on notions of anticipation and return, which readers must adopt as practices in order to make sense of the seemingly incongruent picture of incidents depicted across the city.

Joyce's novel demands its readers keep in mind a staggering array of events, details, and memories as they move forward in the text, and this requirement befits an apprehension of the diverse and variable collective urban life that Joyce's Dubliners embody. Similar to Kenner and Howes, Michael Rubenstein focuses on the novel's 'intercutting of primary narrated events with repetitions of previously narrated events,' but he understands Joyce's technique as imagining the type of belonging that Anderson argues print literature—especially the novel and the newspaper—began to foster in the service of the nation in the eighteenth century.[6] According to Rubenstein, '"Wandering Rocks," more than any other episode of *Ulysses*, stages simultaneity; the nineteen sections of the episode narrate moments in the life of the city collected from a disparate range of geographical locations.'[7] This structure provides, as Anderson puts it in his study of nationalism, 'a complex gloss upon the word "meanwhile."'[8] Borrowing a phrase from Walter Benjamin, Anderson describes the nation as being premised on the 'idea of "homogeneous, empty time," in which simultaneity is, as it were, transverse, cross-time, marked not by prefiguring and fulfilment but by temporal coincidence, and measured by clock and calendar.'[9] Even more clearly than the novel, the newspaper demonstrates this notion of simultaneity with its arbitrary 'inclusion and juxtaposition' of events, which 'shows that the linkage between them is imagined' by way of their strictly 'calendrical coincidence.'[10] In 'Wandering Rocks,' Joyce cultivates this very idea of simultaneity, as his characters jointly make their way through 'homogeneous, empty time' despite their separate intentions and distinct experiences. In the formal centre of Joyce's novel lies a chapter that opens up a larger picture of a community in motion; the separate members of that group may not recognise the larger collective experience in which they participate, but the narrative technique of the episode performs that imaginative work for them.

Figural History and Dublin's Political Ghosts

Though the Dublin of *Ulysses* is still under British rule in 1904, 'Wandering Rocks' offers a vision of the city that renders its citizens in the garb of national compatriots: they appear as a disparate yet imaginatively connected community moving through time together. Enda Duffy argues that Joyce's decision to use 1904 as the backdrop for his novel contributes to the seemingly premature national features of the Irish offered up in *Ulysses*:

> By setting his book in a relatively uneventful earlier year, while writing it amid violent, revolutionary and transformative times, *Ulysses* can know the future without admitting to such knowledge. It can secrete within its account of a day in 1904 a palimpsest or secret history of the revolutionary occurrences of 1914–21, when the book was being written. *Ulysses*, then, double-times its readers, encoding a post-1904 future which the author of 1914–21 has seen unfold into a book which appears to be only able to look back.[11]

To see Dublin in *Ulysses* as the capital of an independent nation is not, then, wishful thinking or loose historicising. Instead, as Duffy would have it, readers of the novel see colonial Dublin through the lens of its eventual independence. Like the aesthetics of delay and parallax that Kenner and Howes explain as transforming earlier and later events in Joyce's novel, the temporal setting of the text allows the experiences of both Ireland's colonial subservience and its revolutionary success to overlap and resonate with one another. Readers end up thinking about two periods of Ireland's history while they also move between variant descriptions of the same experiences told at different points in the text. If the tension of the 1904 setting described by Duffy allows Ireland's national future to shade the historical moment of the novel's temporal present, then the reverse is also true. In the picture of Dublin *Ulysses* offers, Joyce cannot look ahead to the soon-to-be capital of a newly independent republic without also glancing back at the long history of failures and betrayals that characterise the relationship between that city's citizens and the British Empire. To remind his post-independence readers that the legacy of empire has a long reach, one which Dublin could not easily cast off, Joyce inscribes an understanding of time in 'Wandering Rocks' that opposes the 'homogeneous, empty time' through which the national community typically travels. Benjamin himself describes a relationship between the past and the present that is 'not temporal in nature but figural.'[12] According to this other system of time, which is so different from the clock and calendar of the nation, an historical moment acquires significance 'posthumously, as it were, through events that may be separated from it by thousands of years.'[13] When two events come together in figural history, the significations between those separate moments are made manifest, even though their connections would normally have been obscured within the linear narratives of the past offered by the nation. To remedy such limited vision in *Ulysses*, Joyce presents a figural history of Dublin's colonial and independent periods alongside his version of 'homogeneous, empty time' in 'Wandering Rocks.'

Before the rise of nation states and the forms of imagining that they require, figural time structured and supported a sense of belonging in premodern religious communities and dynastic realms—authoritarian social structures more in line with a colonial city like Dublin than a democratic national capital would be.[14] Discussing the figural imagination of the Church Fathers, Erich Auerbach explains how '[f]igural interpretation establishes a connection between two events or persons, the first of which signifies not only itself but also the second, while the second encompasses or fulfils the first.'[15] This approach differs from an allegorical mode of interpretation, as a figural event does not signify some idea or virtue or emotion: it signifies some other definite event. In figural history, the past and present contain entirely real and material separate events, on the one hand, and they are also, on the other, fully actualised in a simultaneous moment connecting the two events through a conception of history radically different from the horizontal progress of 'homogeneous, empty time.' Figuralism recognises the same content being encompassed by an early event and a later one, by a model and a copy, by some original and its ghost image.

Just as historical events can be separated by thousands of years in a figural understanding of history, one must wait several, sometimes hundreds, of pages for the significance of an event to become apparent in *Ulysses*. It is, therefore, no surprise that in such a text there exists a character that is a figure for some other moment that is promised and not yet present. Charles Stewart Parnell, like the ghost of Hamlet's father described by Stephen in the National Library, is 'a king and no king' waiting for redemption in a new Ireland, and his ghost returns to Dublin in Joyce's text.[16] Parnell's legacy comes up elsewhere in Joyce's fiction, but neither in *Dubliners* nor in *A Portrait of the Artist as a Young Man* does that ghost return to Dublin in a way that fulfils the role that Parnell prefigured for Ireland. In 'Ivy Day in the Committee Room,' ivy leaves in the men's lapels bear witness to his presence in Dublin's public memory, and Mr. Hynes's 'piece' on Parnell's death predicts his rising from the dead.[17] However, as in other stories from *Dubliners*, these characters are ossified and habitual in their thinking of the past and of Parnell, and their submission to the new English king's visit to Ireland signals Parnell's harmlessness as a spectre at that moment.[18] In *Portrait*, Parnell's legacy looms over the Christmas dinner scene, and there he proves to be a much more disruptive shade, but any formidable engagement with the promise of Parnell's prefigured role in Ireland on the part of Stephen will come only through his reinterpretation of heretical language—'the language he heard against God and religion and priests in his own home [...], the language with which the priests and priests' pawns broke Parnell's heart and hounded him into his grave.'[19] This reinterpretation will enact a refiguring not of Parnell the betrayed nationalist leader, but of Stephen as the artist that will not serve. The afterlife of Ireland's 'king and no king' is much more limited in these works than it is in *Ulysses*, where the ghost of Parnell returns to Dublin and signals the two historical periods on either side of 1904 that Joyce's Irish experience simultaneously in that city.

Parnell's Political and Spectral Returns

As leader of the Irish Parliamentary Party, Parnell 'brought Ireland to the threshold of statehood under home rule,' according to Frank Callanan, but he could not bring it any farther than that.[20] The divorce granted to Captain W. H. O'Shea on 17 November 1890, and the evidence it leveled against his wife and Parnell, brought an end to the latter's leadership of the Irish party. With courageous resolve, he campaigned throughout Ireland to re-establish his position, but the physically harried Parnell would not finish this 'punishing railway odyssey' before dying on 6 October 1891.[21] Joyce describes these last days in 'The Shade of Parnell': 'He went from county to county, from city to city, "like a hunted hind", a spectral figure with the signs of death upon his brow. Within a year he died of a broken heart at the age of forty-five.'[22] Besides suffering from a broken heart, rheumatism kept Parnell in persistent pain while travelling around the country. A bandaged arm that hung in a sling and a weakened grip when shaking hands, if he was able to shake hands at all, were but two of the many signs of his sickness. Parnell appeared at convention after convention, but he appeared 'wretchedly ill.'[23]

Standing in the rain at Creggs, Parnell could not speak of his political return without admitting that his health had declined:

> Physically I am not in a good condition to address you today. Nothing but a desire not to disappoint the men that I see around me overcame the orders of my doctor that I was to go to bed last night when I arrived in Dublin. However, I do not think that any very material damage will come to me from this meeting. If I was to allow the suggestion of such a thought we should have our enemies throwing up their hats and announcing that I was buried before I was dead [...]. I have to tell you that you must not take me today as I always am, and I hope when I see you again at the next general election, and to ask for your votes, that you will find a very considerable improvement physically (*cheers*).[24]

As in Joyce's depiction of this campaign, there is a phantasmal quality attached to Parnell's own description of himself. His insistence that a premature burial will not keep him from returning at the next election eerily foreshadows the myths that Bloom and other funeral-goers ponder in the 'Hades' episode of *Ulysses*: 'Some say he is not in the grave at all. That the coffin was filled with stones. That one day he will come again.'[25] Alfred Robbins, writer for the *Birmingham Daily Post* during Parnell's last year, suggests that Parnell's death was itself a spectral return to the history of Irish politics:

> On the night of September 30th, he left Ireland for the last time, refusing to heed Kenny's assurance that he was unfit to travel. 'I shall be all right,' said Parnell; 'I shall come back on Saturday week.' These were his last words on Irish soil; and the pledge was kept. On the promised Saturday week, October 10th, he came back to Ireland, to be laid at rest at Glasnevin, the Dublin cemetery in which Daniel O'Connell, his greatest predecessor in Irish leadership, likewise sleeps.[26]

Parnell's fierce effort to return to his role as leader of the Irish party becomes figured in Robbins's words, in the myths recalled by the funeral-goers of the 'Hades' episode, and in Parnell's own self-description as a persistence that lasts beyond the grave. Bloom debunks this posthumous return and sardonically advises against it: 'Highly unlikely of course there was even a shadow of truth in the stones and, even supposing, he thought a return highly inadvisable, all things considered.'[27] In the above descriptions of Parnell's last days, however, there is a suggestion that the parliamentary leader would not rest, even in death, but would continue to find a way back into Irish politics. Depictions of his tolling campaign slide into macabre imaginings that the dead Parnell would continue to haunt an Ireland in which there was still unfinished nationalist business.

In 1904 Dublin, there is another reason for the broken-hearted ghost of Parnell to be restless: when the funeral-goers in 'Hades' pass the site of Parnell's monument, it is still unfinished.[28] In Bloom's mind, this structure's bare base and Parnell's political fall become linked: 'Foundation stone for Parnell. Breakdown. Heart.'[29] If the public memorial for Parnell is still incomplete in 1904, then his afterlife in the city is not officially situated by a designated place of hallowed ground. Instead, the 'king and no king' of Ireland haunts the city through a dislocated public memory, one which circulates through myths, rumours, and ghost stories.[30] In 'Wandering Rocks,' where the narration follows a number of Dubliners, the appearance of Parnell's ghost fills the city with Ireland's unsettled history. Each moment is linked back to the figure of Parnell, and from him to the unfinished business of Ireland's struggle for home rule, and still further to the Irish experience of colonial oppression. The relationship between these events is not temporal but figural, and so *Ulysses* suggests that Ireland's history of colonialism and Parnell's struggle has not yet been redeemed fully. The shade of Parnell is a figure for that future day 'when the new Ireland soon enters into the palace' of national statehood, but the Dublin he haunts in *Ulysses* is always also the colonial city of Ireland's past.[31] Both times are, this ghost reminds us, incompletely understood without their being seen in the shadow of each other.

The Fulfilment of Parnell's Spectral Return to Dublin

'Coming events cast their shadows before,' Bloom tells us.[32] Rightly then, *Ulysses* suggests the possibility of Parnell's apparitional return to Dublin even before the 'Wandering Rocks' episode enacts that moment of historical metempsychosis. In 'Lestrygonians,' Bloom notices John Howard Parnell—the late Parnell's brother—passing through the street: 'There he is: the brother. Image of him. Haunting face. Now that's a coincidence.'[33] This sighting is coincidental in Bloom's mind because he has just been thinking about how Charles Parnell nourished a sense of fascination among the Dublin crowds.[34] Now upon seeing John Howard Parnell, Bloom feels a sense of wonderment: the abrupt pause in his thoughts suggests that he believes he has seen Charles Parnell ('There he is') but then realises that it is only 'the brother.'[35] John Howard is apparently the spitting 'Image of' Charles Stewart, but there is, nonetheless, something 'Haunting' about John Howard's appearance.[36] These two men become linked in a spectral way for Bloom. They are not just images of but figures for one another: 'Great man's brother: his brother's brother.'[37] It is unclear which Parnell is more real, which is living and which is dead: John Parnell looks as though he has eaten a bad egg, and so Bloom makes the pun, 'Poached eyes on a ghost.'[38] It is as if Charles Parnell's spectre has suddenly transmigrated into the body of his living brother. Maybe that which Simon Dedalus has said about John Howard is true: 'when they put him in parliament that Parnell would come back from the grave and lead him out of the house of commons by the arm.'[39] For a second, Bloom thinks that this ghost indeed has come again.

The ghostly imagery surrounding John Howard Parnell in 'Lestrygonians' works its way into 'Wandering Rocks.' In section eight of the latter episode, Ned Lambert shows the reverend Hugh C. Love 'the most historic spot in all of Dublin.'[40] This spot, however, is an ostensible tomb for the ghosts of Ireland's past:

The vesta in the clergyman's uplifted hand consumed itself in a long soft flame and was let fall. At their feet its red speck died: and mouldy air closed round them.

—How interesting! A refined accent said in the gloom.

—Yes, sir, Ned Lambert said heartily. We are standing in the historic council chamber of saint Mary's abbey where silken Thomas proclaimed himself a rebel in 1534.[41]

At best, revolution has become little more than local colour in this scene. Even the interest shown by the clergyman, whom Lambert describes as 'well up in history,' cannot bring the Irish out of the torpor of their colonial experience.[42] Amidst this uncritical contentedness with the past, Joyce places an interpolation. While the clergyman muses over the best angle for his camera shot, a phantasmal line of text interrupts the scene and inserts the image of John Parnell's face:

In the still faint light he moved about, tapping with his lath the piled seedbags and points of vantage on the floor.

From a long face a beard and gaze hung on a chessboard.

I'm deeply obliged, Mr. Lambert, the clergyman said. I won't trespass on your valuable time...[43]

At this point in 'Wandering Rocks,' the identity of the intrusive face is unknown, but Joyce has provided a clue earlier in the novel. When Bloom catches sight of John Howard Parnell in 'Lestrygonians,' he surmises that the great man's brother will 'Drop into the D.B.C. probably for his coffee, play chess there.'[44] In section 16 of 'Wandering Rocks,' Bloom's guess is substantiated, as Joyce fleshes out the figural relationship between Charles Stewart and John Howard Parnell, who in fact turns up at the Dublin Bread Company at that point in the episode.

The prefigured return of Parnell comes about when Buck Mulligan and the Englishman Haines walk into the D.B.C. and espy the great Parnell's sibling:

—Parnell's brother. There in the corner.

They chose a small table near the window, opposite a longfaced man whose beard and gaze hung intently down on a chessboard.

—Is that he? Haines asked, twisting round in his seat.

—Yes, Mulligan said. That's John Howard, his brother, our city marshal.

> John Howard Parnell translated a white bishop quietly and his grey claw went up again to his forehead where it rested. An instant after, under its screen, his eyes looked quickly, ghostbright, at his foe and fell once more upon a working corner.[45]

Just as Bloom remembers Charles Parnell's having 'used men as pawns' politically, John Howard Parnell now moves his chess pieces across the board.[46] This game takes the place of the corporation meeting that Bloom in 'Lestrygonians' remembers as being scheduled today.[47] The description of John Parnell's hand as a 'grey claw' also signals his brother, whose rheumatic arm and hand pained him during his last days moving from town to town in a last gasp to revive his career. Simon Dedalus's prophecy was close to the mark: the ghost of Parnell has not led John Howard out of the house of commons, but a 'ghostbright' vision has seated the city marshal at a table in the D.B.C. The textual interpolations that connect this scene to other parts of the episode suggest that the doubled presence of the shade of Parnell and his haunted brother serve a critical role in Dublin today. The intrusion of John Howard's haunted face upon the council chamber of Saint Mary's abbey provides a jolt to the stasis of that scene. Furthermore, John Howard does not acknowledge the viceregal cavalcade when it rolls through section 19 and puts on display Britain's privileged position over the city: 'From the window of the D.B.C. Buck Mulligan gaily, and Haines gravely, gazed down on the viceregal equipage over the shoulders of eager guests, whose mass of forms darkened the chessboard whereon John Howard Parnell looked intently.'[48] Those in the D.B.C. who do watch the cavalcade's procession throw a shadow across the chessboard, suggesting that the shade of Parnell is with his brother, resisting the panoptic gaze of the city's imperial occupiers.

The haunted face of John Howard Parnell in 'Lestrygonians' foretells the later appearance of his brother's ghost, and in 'Wandering Rocks' that shade returns to Dublin to mark that city as a place where periods of history overlap. This suggests that there are multiple ways for Joyce's Dubliners to locate themselves in relation to the past. In true figural fashion, the ghost of Parnell does not fully emerge until later parts of 'Wandering Rocks' substantiate earlier ones. To say that this spectre first appears at some specific moment or in one particular section of the episode is to ignore the full effect of the text's multiple interpolations: when Parnell's ghost haunts one section of the episode, it haunts every other section, too. The final section of 'Wandering Rocks,' in which the viceregal cavalcade moves throughout the city, attempts to offset the technique of these textual interpolations. Many of the Dubliners from earlier in the episode witness the imperial equipage as it progresses in a triumphant procession. This linear narrative of the city, told from an empowered perspective, sets the previous 18 sections off as a phantasmal mirror image of this final scene. It attempts to subsume these heterogeneous Dublin experiences within 'homogeneous, empty time.' Fredric Jameson has noted that the presence of the viceregal cavalcade in Dublin demonstrates how history is 'already part of the urban fabric' of this community.[49] It is perfectly natural for this official procession and sign of Ireland's colonial legacy to make its way through the city's streets. The final section of 'Wandering Rocks,' however, also suggests that there are fissures in the urban fabric that break up the equipage's linear progression through 'homogeneous, empty time.' The carriages are 'unsaluted' by Mr. Dudley White, and the hands of Blazes Boylan 'forgot to salute'; other gestures of recognition remain 'unperceived.'[50] Even the natural fabric of the city itself offers a sarcastic sign of resistance: 'From its sluice in Wood quay wall under Tom Devan's office Poddle river hung out in fealty a tongue of liquid sewage.'[51] Parnell's spectre links these cracks in the official narrative describing the viceroy's procession to an oppositional picture of Dublin's colonial experience. The ghost of Parnell counters the cavalcade's imperial threat to Dublin by linking the present moment with the struggle for home rule and with the promise of Ireland's ultimately successful independence. Through Parnell's ghost the past is related figurally to a future moment of fulfilment, and so Joyce's depiction of Dublin in 1904 becomes a moment torn apart and discontinuous with the ordinary flow of time.

The Dislodged Time of 'Wandering Rocks' and Dublin's Communal Crisis

'Wandering Rocks' is also a ghost story within the novel's Odyssean parallel. Odysseus never passed through the Wandering Rocks, opting instead to travel between Scylla and Charybdis. Joyce doubles back and gives this tale, belonging properly to Jason and the Argonauts, a second life. As an instance of recapitulation or redeployment, 'Wandering Rocks' can thus be understood as a beginning in the sense described by Edward Said: 'a beginning intends meaning, but the continuities and methods developing from it are generally *orders* of *dispersion*, of *adjacency*, and of *complementarity*.'[52] Meaning in this episode does not arise from a privileged centre dominating all that derives from it; 'Wandering Rocks' is not, in Said's terms, an origin. The disrupted and interconnected structure of 'Wandering Rocks' reminds us that Joyce's Dublin coheres through an order of dispersion and adjacency. Fittingly, the action in this episode does not centre on either of the two characters that have been the novel's chief concern up until this point. After the chance encounter between Stephen and Bloom at the end of 'Scylla and Charybdis,' the novel splinters and follows a number of other Dubliners throughout the city. Instead of realising Stephen's desire to become consubstantial with the father, Joyce's narrative turns constellational and represents a social rather than self-created history: at this point, *Ulysses* foregrounds the density of interpersonal, public, and political histories in the city. The moment in which Bloom, the figure of the exiled Wandering Jew, passes between Stephen and Buck Mulligan on the Library steps becomes the moment in which all of the other Dubliners are also cast as exiles in their own country. The main character of 'Wandering Rocks' is the entire city of Dublin. As part of a colony, however, Dublin is always culturally adjacent to London; it is the capital of a nation almost. The public history and the personal memories evoked by the architecture and material presence of their occupied city—such as the tombstones of failed Irish leaders in the Glasnevin cemetery where Paddy Dignam is interred or the viceregal cavalcade that closes the 'Wandering Rocks' chapter—remind these Dubliners of their lives being rooted in the disadvantageous end of the relationship empire creates between colonial markets and imperial capital. The peripheral status preventing Ireland from being a nation outright figures time in Joyce's Dublin as a collective experience that is always somewhat dislodged from the measures used to coordinate urban life elsewhere.

Joyce also performs a recapitulation of the novel's formal structure through this episode. Placed in the centre of the novel's 18 episodes, the 19 sections of 'Wandering Rocks' serve as a microcosm of the whole text. *Ulysses* begins in 'Telemachus' with Buck Mulligan's mocking the mass, turning that sacrament into a comedic stage show; similarly the first section of 'Wandering Rocks' follows Father John Conmee who, reading the office, transforms into the mock-Byronic 'Don John Conmee.'[53] Apparently, in 'Wandering Rocks,' Joyce's novel starts over. With this focus on Father Conmee in the first section, and a camera-like eye on the viceregal cavalcade in the final section, representatives of Stephen's two masters, one Roman and one British, function as reminders of Ireland's colonial history. At the structural heart of *Ulysses*, framed between the coloniser against whom he fought and the Church that brought him low, the ghost of Charles Stewart Parnell returns to Dublin with the promise of a new history.

In 1904 Dublin, the problem of history is shared by many; Joyce represents time in 'Wandering Rocks' as a communal crisis, a dispersed order, a multiply ruptured text. The experience of colonial oppression links all the characters in *Ulysses* to one another and also to the past, especially to the shade of Parnell. Those in the present have, in the words of Benjamin, 'a *weak* Messianic power' to which that ghost has a claim.[54] After all, they were the ones who broke his heart:

> The shade of the 'uncrowned king' will weigh upon the hearts of those who remember him, when the new Ireland soon enters into the palace […] but it will not be a vindictive shade. The sadness that devastated his soul was, perhaps, the profound conviction that, in his hour of need, one of the disciples who had dipped his hand into the bowl with him was about to betray him.[55]

The presence of Parnell's ghost in Dublin in 'Wandering Rocks' represents a moment of critical re-evaluation. This ghost has not returned to damn the living—it is not 'a vindictive shade'—but rather, he comes to make Ireland look at its colonial past and question whether there are not more ways out of that labyrinth yet to be found. Joyce, in the end, never really knew Dublin as anything other than a colonial city; once Home Rule came to Ireland, its capital always required some marvellous work of the imagination for it to take shape in the artist's mind.

Notes

1. James Joyce, *Ulysses*, ed. Hans Walter Gabler (New York: Vintage, 1986), p. 91.
2. Benedict Anderson, *Imagined Communities* (New York: Verso, 1991), p. 26.
3. See James Hansen, 'The Uncreating Conscience: Memory and Apparitions in Joyce and Benjamin,' *Mosaic* 34, no. 4 (December 2001): pp. 85–106.
4. Hugh Kenner, *Ulysses* (Baltimore: Johns Hopkins UP, 1987), p. 81.
5. Marjorie Howes, 'Memory: "Sirens,"' in *The Cambridge Companion to Ulysses*, ed. Sean Latham (Cambridge: Cambridge UP, 2014), p. 129.
6. Michael Rubenstein, 'City Circuits: "Aeolus" and "Wandering Rocks,"' in *The Cambridge Companion to Ulysses*, ed. Sean Latham (Cambridge: Cambridge UP, 2014), p. 121.
7. Rubenstein, 'City Circuits,' p. 117.

8. Anderson, *Imagined Communities*, p. 25.
9. Anderson, *Imagined Communities*, p. 24.
10. Anderson, *Imagined Communities*, p. 33.
11. Enda Duffy, 'Setting: Dublin 1904/1922,' in *The Cambridge Companion to Ulysses*, ed. Sean Latham (Cambridge: Cambridge UP, 2014), p. 81.
12. Walter Benjamin, *The Arcades Project*, eds. Howard Eiland and Kevin McLaughlin (Cambridge: Harvard UP, 1999), p. 463.
13. Walter Benjamin, 'Theses on the Philosophy of History,' in *Illuminations*, trans. Harry Zohn (New York: Schocken, 1968), p. 263.
14. For a discussion of the transformation of the temporal imagination of religious communities and dynastic realms to that of nations, see Anderson, *Imagined Communities*, pp. 22–24.
15. Erich Auerbach, 'Figura,' in *Scenes from the Drama of European Literature* (Minneapolis: Minnesota UP, 1984), p. 53.
16. Joyce, *Ulysses*, p. 155.
17. James Joyce, *Dubliners* (New York: Penguin, 1992), p. 119, 131.
18. Discussing the politician's 'political manner and method,' Frank Callanan argues that Parnell's own 'taciturn elusiveness within his own myth was to become a defining feature of Parnell's spectral presence in Joyce's writing,' particularly in 'Ivy Day in the Committee Room.' See Frank Callanan, 'The Parnellism of James Joyce: "Ivy Day in the Committee Room,"' *Joyce Studies Annual* (2015): p. 88.
19. James Joyce, *A Portrait of the Artist* (New York: Penguin, 1993), p. 33.
20. Frank Callanan, *The Parnell Split: 1890–91* (Syracuse: Syracuse UP, 1992), p. 1.
21. Callanan, *The Parnell Split: 1890–91*, p. 179.
22. James, Joyce, 'The Shade of Parnell,' in *Occasional, Critical, and Political Writings*, ed. Kevin Barry (Oxford: Oxford UP, 2000), p. 196.
23. Callanan, *The Parnell Split: 1890–91*, p. 179.
24. From the *Freeman's Journal*, 28 September 1891 (cited in Callanan, *The Parnell Split: 1890–91*, p. 180).
25. Joyce, *Ulysses*, p. 93.
26. Sir Alfred Robbins, *Parnell: The Last Five Years* (London: Thornton Butterworth, 1926), p. 194.
27. Joyce, *Ulysses*, p. 530.
28. For more details on the Parnell monument, and its eventual completion in 1911, see Don Gifford and Robert J. Seidman, *Ulysses Annotated* (Berkeley: University of California Press, 1988), p. 111.
29. Joyce, *Ulysses*, p. 79.
30. Luke Gibbons has described a similarly dislocated, oppositional public memory that is 'graphically demonstrated in *Ulysses* by the image of the "five tallwhitehatted sandwich men" passing by the slab at the corner of St.

Stephen's Green "where Wolfe Tone's statue was not."' See Luke Gibbons, '"Where Wolfe Tone's statue was not": Joyce, monuments and memory,' in *History and Memory in Modern Ireland*, ed. Ian McBride (Cambridge: Cambridge UP, 2001), pp. 140–141.

31. Joyce, 'The Shade of Parnell,' p. 196.
32. Joyce, *Ulysses*, p. 135.
33. Ibid.
34. Joyce, *Ulysses*, p. 134.
35. Joyce, *Ulysses*, p. 135.
36. Ibid.
37. Ibid.
38. Ibid.
39. Ibid.
40. Joyce, *Ulysses*, p. 189.
41. Ibid.
42. Joyce, *Ulysses*, p. 190.
43. Ibid.
44. Joyce, *Ulysses*, p. 135.
45. Joyce, *Ulysses*, p. 204.
46. Joyce, *Ulysses*, p. 135.
47. Ibid.
48. Joyce, *Ulysses*, p. 208.
49. Fredric Jameson, 'Modernism and Imperialism,' in *Nationalism, Colonialism, and Literature*, ed. Seamus Deane (Minneapolis: University of Minnesota Press, 1990), p. 63.
50. Joyce, *Ulysses*, pp. 207–208.
51. Joyce, *Ulysses*, p. 207.
52. Edward W. Said, *Beginnings: Intention and Method* (New York: Columbia UP, 1985), p. 373. Emphasis in the original.
53. Joyce, *Ulysses*, p. 184.
54. Benjamin, 'Theses on the Philosophy of History,' p. 254. Emphasis in the original.
55. Joyce, 'The Shade of Parnell,' p. 196.

References

Anderson, Benedict. *Imagined Communities*. New York: Verso, 1991.

Auerbach, Erich. 'Figura.' In *Scenes from the Drama of European Literature*, 11–76. Minneapolis: University of Minnesota Press, 1984.

Benjamin, Walter. 'Theses on the Philosophy of History.' In *Illuminations*, trans. Harry Zohn, 253–264. New York: Schocken, 1968.

———. *The Arcades Project*. Edited by Howard Eiland and Kevin McLaughlin. Cambridge, MA; London, UK: Harvard University Press, 1999.

Callanan, Frank. *The Parnell Split: 1890–91*. Syracuse: Syracuse University Press, 1992.

———. 'The Parnellism of James Joyce: "Ivy Day in the Committee Room".' *Joyce Studies Annual* (2015): 73–97.

Duffy, Enda. 'Setting: Dublin 1904/1922.' In *The Cambridge Companion to Ulysses*, ed. Sean Latham, 81–94. Cambridge: Cambridge UP, 2014.

Gibbons, Luke. '"Where Wolfe Tone's Statue Was Not": Joyce, Monuments and Memory.' In *History and Memory in Modern Ireland*, ed. Ian McBride, 139–159. Cambridge: Cambridge UP, 2001.

Gifford, Don, and Robert J. Seidman. *Ulysses Annotated*. Berkeley: University of California Press, 1988.

Hansen, James. 'The Uncreating Conscience: Memory and Apparitions in Joyce and Benjamin.' *Mosaic* 34, no. 4 (December 2001): 85–106.

Howes, Marjorie. 'Memory: "Sirens".' In *The Cambridge Companion to Ulysses*, ed. Sean Latham, 128–139. Cambridge: Cambridge UP, 2014.

Jameson, Fredric. 'Modernism and Imperialism.' In *Nationalism, Colonialism, and Literature*, ed. Seamus Deane, 43–66. Minneapolis: University of Minnesota Press, 1990.

Joyce, James. *Ulysses*. Edited by Hans Walter Gabler. New York: Vintage, 1986.

———. *Dubliners*. New York: Penguin, 1992.

———. *A Portrait of the Artist*. New York: Penguin, 1993.

———. 'The Shade of Parnell.' In *Occasional, Critical, and Political Writings*, ed. Kevin Barry, 191–196. Oxford: Oxford UP, 2000.

Kenner, Hugh. *Ulysses*. Baltimore: Johns Hopkins University Press, 1987.

Robbins, Sir Alfred. *Parnell: The Last Five Years*. London: Thornton Butterworth, 1926.

Rubenstein, Michael. 'City Circuits: "Aeolus" and "Wandering Rocks".' In *The Cambridge Companion to Ulysses*, ed. Sean Latham, 113–127. Cambridge: Cambridge University Press, 2014.

Said, Edward W. *Beginnings: Intention and Method*. New York: Columbia University Press, 1985.

CHAPTER 11

'It's only history': Belfast in Rosemary Jenkinson's Short Fiction

Dawn Miranda Sherratt-Bado

Twenty years after the signing of the Good Friday/Belfast Agreement in 1998, this chapter analyses the ways in which this event impacts upon conceptions of Belfast, history, and identity in Rosemary Jenkinson's short story collections *Contemporary Problems Nos. 53 & 54* (2004) and *Aphrodite's Kiss and Further Stories* (2015). Jenkinson is an acclaimed playwright, but she has been 'writing stories for much longer' and her short fiction remains underexplored.[1] She trains her dramatist's eye on post-Agreement Belfast within her stories, which feature highly pressurised interactions in confined city spaces: Belfast's 'walled communities,' pubs, and 'close streets cosseted by hills.'[2] These moments have the potency of scenes from a play, and their rich local imagery and lively use of dialect lend a sense of immediacy. The compressed short story form also enhances the potential for an explosive effect. Jenkinson problematises notions of contemporary Belfast as a 'post-conflict' space by exposing entrenched sociopolitical tensions and considering how these inflect exchanges between locals and tourists, as well as with the city itself. Her portrayal of the contemporary city also functions as a commentary on the commercialisation of Belfast and its history. The economic subtext of the Agreement signals a break with the city's 'troubled' past in order to align

D. M. Sherratt-Bado (✉)
Independent Scholar, Maynooth, Ireland

M. Beville, D. Flynn (eds.), *Irish Urban Fictions*, Literary Urban Studies, https://doi.org/10.1007/978-3-319-98322-6_11

with a global capitalist future. Therefore the 'new' Belfast is circumscribed by its own corporatised, 'post-conflict' image in a process which is paradoxically violent; for the progressivist discourse of the Agreement dismisses fraught identitarian narratives as anachronistic. In Jenkinson's tales this disjuncture manifests as a crisis of narrative, and her characters remain adrift. She emphasises the complexities of Northern Irish identity in her Belfast stories, thereby reasserting the local within a culture that has become 'globally entangled.'[3]

Often the prevalent theme in Jenkinson's stories is alienation, for her characters experience Belfast as 'a place apart'—a frequent trope in Northern Irish literature.[4] This effect is compounded by the fact that, as Colin Graham contends, 'peace was bought at the price of dissociation rather than consociation' via the delimiting terms of the Agreement, 'in which the political process drifts further off from the everyday, and in which there is a structural necessity for cultural identities to be put in abeyance.'[5] The Agreement reiterates the unresolved sectarianism that has historically underpinned conflict in the North, whilst simultaneously signalling an end to this conflict. Correspondingly, Jenkinson remarks, 'the best thing about using Belfast as a backdrop is that the city is a threat—it can't help it because of its past. What other city would be overlooked by a hill as menacingly named as Black Mountain?'[6] In Jenkinson's stories Black Mountain metaphorises the hulking threat of 'the historical detritus left by the Troubles,' which the current dispensation endeavours to divert, but which continues to pile upon itself in a colossal heap.[7] In her collections, she portrays characters who return to Belfast after the Agreement and who move through the cityscape under the shadow of its sublimated history.

'Taking Care of Business'

Jenkinson returned to Belfast in 2002, four years after the Agreement and sixteen years after she moved to England. She states, 'I remember the "Big Return" at Aldergrove International Airport even now—the homely whiff of slurry that always hits you as you step off the plane, and the huge Harp sign on the terminal, and how thrilled I was to be back but (and there's always the "but") my parents and brother didn't live there any more.'[8] Jenkinson published her debut short story collection, *Contemporary Problems Nos. 53 & 54*, in 2004 and the book explores this condition of unhomeliness which characterises post-Agreement Belfast. Part II of the collection, which I will term the Belfast sequence, is a set of four stories that reflect the contemporaneous cityscape in which they were written, between 2002 and 2003. The protagonists in 'Taking Care of Business' and 'A City Loved' have returned recently to Belfast, where they encounter a fragile peace. In 'Taking Care of Business,' Pat McKelvey comes back to settle her brother's affairs after his death. Initially Pat has a rosy view of her hometown: she 'watched the golden glow of the setting sun light up the shallow city basin within the hills, filling it like a shining dustbowl, silhouetting the far-off steeples of West Belfast. It was worth coming back for this view alone.'[9] Pat arrives in August 2002 to a supposedly 'post-conflict' Belfast; however, she soon finds herself amidst the chaos of the Sandy Row riots.[10] The story opens with the depiction of a city in disorder:

> On a hot Sunday evening in August, Sandy Row was hiving. Youths were hanging round the street corners in groups, fists in the pockets of their hooded shirts, still clenched from the previous night's riots. They yawned from sleeping all day. Outside the Royal, a group of men, golded-up in their chains, were drinking carry-outs, using the bin as a table. Nearby, children … were swigging from a bottle behind a knocked-down wall that still bore vestiges of an old UVF mural.[11]

In this scene, Jenkinson illustrates the residual sectarianism and societal unrest that belie 'the putative switch from the Troubles to the Peace.'[12] She alternates Pat's view of the disturbances as a recent returnee with those of Merv and Hoey, local youths who have never left the city. Merv is enrolled in a programme at the Ulster Society for the Promotion of British Heritage and Culture. Jenkinson writes, 'They passed the blue shop front of The Ulster Society. Merv attended a class there on a Tuesday night to learn about politics and the art of debating … it was a take-on, just a way of trying to keep people off the streets, giving them false hopes that would never come to anything.'[13] When Merv is not in class, he continues to roam the streets with Hoey: 'No one took any notice of the two youths, least of all the tall Canadian with the camcorder. His friends felt uncomfortable and decided to go back to the hostel but they left him their camera. "I'll get you some good shots," he promised them, slinging the camera across his shoulder like an ammo belt.'[14] In a recent interview,

Jenkinson comments that 'tourists seem to think the Troubles are totally over but that's not the case.'[15] In 2002, when the story is set, a series of real-life riots occurred in Belfast when locals clashed with the PSNI and the British Army.[16]

When news of the previous night's rioting reaches Pat, she walks down to Sandy Row to get the details from the local shopkeeper:

> the only traces of the night's trouble were some telltale petrol slicks on the road, a molten shop sign, boarded shop windows and some ploughed-up earth in the pavement where saplings had been uprooted. But what was left of the buzz—no one could sweep that up, and the greengrocer had been quick to point the finger: 'Sure, it was all because the peelers had bate the bollocks out of a couple of wee young fellas,' he'd said.[17]

Sensing that tensions will mount again that evening, Pat decides to return to her brother's house for safety: 'She was glad that over in England her daughter would have no idea about the riots. She sank into wondering how her life would have been if she'd never been forced out of Belfast.'[18] She is startled from her reverie by the sound of a British Army Chinook. She listens as '[t]he helicopter swerved from its depths like a smoked-out wasp. There was an excited hysteria in her heart as there always was when she viewed something apocalyptic.'[19] Pat goes closer to the window, where she witnesses Merv and Hoey throwing a petrol bomb at the house: 'The glass shattered and Merv lit the rag at the top of the bottle and threw it, already turning and running, almost propelled by the smash and roar behind him.'[20] Pat runs outside, choking on the fumes and 'watch[ing] the hot smoke seep from the mouth of the letter box. An uncomprehending anger seized hold of her. She'd been on the blower to Peter Rea the day after her brother had died and he'd negotiated on her behalf. They'd told him it was all forgotten in these days of amnesty, that she was safe to visit, that the long exile was rescinded.'[21] Pat is outraged by Peter's insistence that it is safe to return to post-Agreement Belfast, but she is even angrier at herself for believing him.

Merv and Hoey escape Pat's neighbourhood in Boyne Court unscathed and head west, where 'about fifty rioters had converged at the end … of the Donegall Road.'[22] There they mill about with the rioters until they see someone 'unleashing a petrol bomb at a Saxon' and 'the police in riot gear' begin 'pouring into the streets.'[23] The crowd disperses into the nearby estates and Merv and Hoey run down an alleyway. Looking around nervously, Hoey observes that in the shop window 'there was a brief notice that stated: "Anyone involved in anti-social activity will be dealt with." Hoey's mouth had gone dry. It had been Merv's idea ... Merv was talking too much bull now.'[24] Hoey and Merv continue running until they encounter the Canadian tourist who is filming the riots with his borrowed camcorder:

> Hoey laid into him, uncertain of how much damage his trainers could do. Merv saw to the camcorder. Words he was learning to bandy about at The Ulster Society suddenly surged into his head in a harsh jumble with each separate kick, words that had recently become a conduit for his rage and hatred: condone, infringement, civil liberty, justifiable [...].[25]

Merv is infuriated by the Canadian tourist's filming of the riots as though they are a spectacle for his consumption. Seduced by scopic terror, the tourist intends to bring home the videotape as a souvenir for himself and his friends—'I'll get you some good shots.'[26] To record the conflict is to appropriate it, and the tape will serve as proof of the Canadian's daring in choosing to stay on the streets while his friends 'decided to go back to the hostel.' When Merv destroys the camcorder, he also experiences an internal assault on the senses whereby 'civic-mindedness' takes on a different significance. His inarticulable rage erupts in a stream-of-consciousness barrage of words that have become decontextualised and therefore meaningless: 'condone, infringement, civil liberty, justifiable.' This is the language of 'Peace Process-speak,' which the Ulster Society imparts to local youths in night classes on 'politics and debate.'[27] In personal correspondence Jenkinson explains, 'That's very much the sort of language which is designed to control you. This story reflects the fact that it is possible to be given all of that jargon and to still feel sectarian hatred. Spouting on about "justice" and "peace building" has not taken away sectarianism in this city. There is an irony here in that we're filled with that sort of language but deep down there is still the potential for violence.'[28] The final sentence of the story ends in ellipses, which indicate a moment of catachresis. Here words fail to denote their proper meaning; however, Merv appears to have no option but to inhabit this discursive space. He attempts to resist his situation via the subversive application of institutional rhetoric. Redeploying the ideological jargon of peace, Merv channels it into an act of violence. He directs his diatribe at an inanimate object—the tourist's smashed camcorder—responding subconsciously to the way in which tourists perpetuate the ubiquitous surveillance culture in Belfast. Furthermore, the final ellipses suggest that 'Taking Care of Business' is a tale without a conclusion. It does not end with 'conflict resolution' or 'reconciliation'—thus, Jenkinson implies that the 'business' of the story remains unfinished.

'A City Loved'

Another story from the Belfast sequence, 'A City Loved,' is focalised through the perspective of Mark, a Catholic from Springfield, a predominantly nationalist Catholic neighbourhood off the Falls Road in West Belfast and a site of continual sectarian violence during the Troubles. Mark experiences this conflict first-hand and it prompts him to leave home: 'There was the time the RUC had got him, hauled him into the back of the Land Rover and drove him to the Protestant side, threatening to chuck him out on the Shankill and shout out, "Taig!"'[29] Mark returns to Belfast after living in Manchester for five years, only to discover that not much has changed in his old neighbourhood. Initially, he finds this cheering; however, 'that was a month or so back and the inner bloom of return had faded and been replaced with a vague sense of foreboding. It wasn't easily explained.'[30] Mark's ambivalence is evident in his description of the variable environment: 'it was the up and down of the drumlins, the stomach-thrill of their steep falls that showed him how happy he was to be back. And the hills dark around Belfast, transforming by a sudden cloud-shift to the light green of a wind-dried stone. Then through the terraced streets of orange brick … the sun running its searchlight rays over every house.'[31] A surveillance culture pervades the shifting, watchful landscape within this story as it traces Mark's movements throughout the city. He is particularly aware of being watched when he crosses into East Belfast to visit his new group of Protestant friends. He remarks, 'Today he was Mark from Springfield. Funny, he even had to introduce himself like that now. Everyone was defined by territory.'[32]

In the working-class residential areas within 'A City Loved,' territory is demarcated by street art. Mark encounters paramilitary murals along the peace walls and he considers their variegated expressions of urban ownership and their relation to the defensive architecture of which they are a part. These murals are self-reflexive texts, as 'words are painted on walls only to be dispossessed and answered on other walls, in other words. Words interrogate the wall. They insinuate themselves into the wall structure like shades.'[33] Accordingly, Mark recounts, 'there was a mural down at the end of his street which said: HISTORY IS WRITTEN BY THE WINNER … he now knew that was false. Under these revisionist skies, history was won by the writers.'[34] Belfast's murals utilise competing and ever-changing iconographic sign systems. However, as Aaron Kelly argues, these icons are 'transformed by … commodity fetishism. Put simply, the image becomes a compensatory fetish object that disavows its actual historical referent and social context. In Walter Benjamin's terms, the image loses its "aura" through both its reification and reproducibility.'[35] Similarly, Liam Kelly maintains, 'the residue of concepts such as "difference" and "logos"' is 'related in an Ulster context to cultural identity, colonialism, text, truth and image.'[36] Mark attempts to discern a readable account of Belfast on its surfaces, one which will provide a truthful historical record. Instead he realises:

> It was a tale of two cities—no, a city of two tales. And it scared him going to the other side with the long legends on low walls, curving away into eternity … But there was one placard he'd seen in the Presbyterian church grounds that lifted his heart and showed him that old beliefs were on the move: LET US FOLLOW AFTER THINGS THAT MAKE PEACE.[37]

Mark wants to believe the promise of peace assured by the Agreement; however, he perceives a parallel between this document and similar guarantees pledged by the Protestant and Catholic Churches, as well as republican paramilitaries: 'You will only be happy in a United Ireland, they said. It was like the promise of eternal life by the Church.'[38] In Mark's purview, these are tautological narratives that enmatrix the subject in a process of foreclosure disguised as progress. The interplay between the surfaces and deep structures of language and of Belfast's architecture within the story enables Jenkinson to analyse the city's paradoxical sense of placelessness.

During his journeys around Belfast, Mark attempts to re-orient himself within a city that has lost its former cultural coordinates. The streetscapes display scrambled messages which are a response to the predominant role of global capitalism in the regeneration of Belfast. As Liam O'Dowd argues, 'What we see are traces of uncertain, tenuous and transient claims on public space—attempts to transform the new space of consumer capitalism into "place" via murals, graffiti.'[39] Mark recalls, 'He'd been across the peace line once in the past month to visit his new friend, Robert, through the scribbled, gaudy streets … Since the Agreement it felt as though the iron curtain had come down with new people coming in and new confidence growing,' and yet more 'peace lines had been erected, more barricades built up.'[40] Similarly, Daniel Jewesbury asserts that '[s] ince 1998, the new Northern Ireland has been undergoing something like the "shock" capitalism foisted on the former communist states.'[41] Mark registers this sense of shock when he journeys into the city centre for a night out with friends. Jenkinson writes, 'He caught the bus into the centre of town, passing the packed restaurants lit up by the ultra-fashionable ultraviolet lights, each window bearing the fiery sunflower faces of people drunk on wine and good food. Stencilled on the pavements in chalk was the Guinness message: BELIEVE IN YOURSELF, with a harp representing the letter V.'[42] Here another form of street art confronts Mark—that of advertising. As he moves from West Belfast into the city centre, sectarian geography gives way to commercial geography. The sidewalk Guinness advertisement echoes Jenkinson's description of being greeted by 'the huge Harp sign on the terminal' in Aldergrove International Airport (renamed Belfast International Airport in 1983). Harp lager is produced by the Guinness Brewery, which merged with Grand Metropolitan in 1997 to form the multinational conglomerate Diageo. The 'huge Harp sign' at Belfast International Airport signifies that Diageo is the gateway to not only (Northern) Ireland, but also the world. Such displays are indicative of 'a new "asectarian" commercialism' which economically and spatially reorganises Belfast city as constitutive of the global marketplace.[43] Diageo explain their name on the corporate website: 'The word Diageo comes from the Latin for day (*dia*) and the Greek for world (*geo*). We take this to mean every day, everywhere, people celebrate with our brands.'[44] In Jenkinson's portrayal of Belfast nightlife, people in the bourgeois city centre are invited to celebrate the ostensible peace and anticipated prosperity by drinking Guinness. Here Northern Ireland is subsumed by Ireland, which is in turn subsumed by Diageo's globality. Through this process of consolidation (Northern) Ireland is paradoxically scattered 'everywhere' and nowhere.[45]

The Guinness promotional message directs the consumer to 'BELIEVE IN YOURSELF, with a harp representing the letter V.' Here the sign of the harp is polyvalent—it is at once a signifier for Guinness stout and for Harp lager, which, as mentioned previously, are produced by the same parent company. It also evokes Irish nationalist iconography, as the harp is a metonym for Ireland. Moreover, the harp is an image 'culled from the excesses of populist versions of restored Irish authenticity.'[46] The promotional slogan is in fact an injunction to 'believe' in this new/old version of (Northern) Irishness which is de-sectarianised and glamorised, its appearance made slick via a process that I have termed elsewhere 'smooth violence.'[47] The 'scribbled' graffiti that Mark observes in the outer, working-class areas of the city are here mimetically 'stencilled' neatly onto the sidewalk.[48] Furthermore, the advertisement's location on the sidewalk can be read as an injunction for the city to believe in itself—as in, to believe in its own mythical self-image. Nevertheless, the fact that the slogan is written in chalk suggests the unsustainability of such a myth. Belfast's corporatising mythos, which is evident in the glossy Good Friday/Belfast Agreement brochure that was distributed to every household in the North, attempts to iron out the 'dialectics of urbanity.'[49] As Aaron Kelly contends, this is due to 'the economic and commercial imperative that underpins not only urban redevelopment but also the Peace Process itself.'[50] He continues, 'The regeneration of Belfast is underscored by a discourse of development' that informs not only debates about the city's economic and spatial arrangement, 'but also an attendant sense of historical development—that the motors of progress are driving the city forwards.'[51] Yet as 'A City Loved' demonstrates, Belfast and its residents are caught in a state of liminality. Mark ponders, 'it was as if he hadn't been able to get home that day a month ago, was still waiting on an isolated stretch somewhere, waiting for a roadblock to be lifted.'[52] This statement illustrates Mark's sense of deep imbrication within the city of Belfast, whose pathways to the future remain enclosed by the terms of the peace process and the Agreement. Mark's sense of perpetual 'waiting' evidences the phenomenon that Richard Kirkland terms 'the vacuum of a lived interregnum.'[53] Kirkland argues that the Northern Irish state remains suspended in a self-perpetuating politico-cultural 'crisis.'[54] Similarly, in Jenkinson's stories, the post-Agreement period is one of non-becoming—it is the lived condition of an eternal present which is yet to be self-determined.

'Scenes from an Empty Attic'

Jenkinson's second book of short fiction, *Aphrodite's Kiss and Further Stories* (2015), also features several Belfast tales. However, unlike *Contemporary Problems*, these seven stories are not presented sequentially. Rather, they bookend the collection, providing a framework for thinking further about the idea of returning 'home' to Belfast. Similarly to Pat in 'Taking Care of Business' and Mark in 'A City Loved,' the protagonists in the two stories that I will examine from this volume have also returned to the city recently. The opening tale, 'Scenes from an Empty Attic,' is a brief, seven-and-a-half-page glimpse into the life of Angelina, a young woman who comes home after teaching abroad, only to lose her way. She recounts, 'During her time away, she'd haunted anyone with an Ulster accent, anyone who'd said "a wee drink", "a wee walk", feeling the pang of the loss. She'd been so happy to come back, but the feeling had faded. Now she was travelling once more, only not across huge spaces but within her own mind.'[55] Unable to establish an individuated self-image, Angelina chases fantasies, similarly to the way in which she chased 'anyone with an Ulster accent' while she was abroad. She spends her nights 'falling and surging on drugs and joy and pain' while partying in tiny derelict flats around Belfast or holed up in Kelly's Cellars.[56] This has been Angelina's nightly routine for the past four years since her return to Belfast, and Jenkinson implies that this behavioural pattern will continue indefinitely.

After spending several years overseas, Angelina is baffled by the massive influx of tourists who flock to post-Agreement Belfast. She frequents Kelly's Cellars, a 'staunchly republican' pub which is now an unexpected tourist attraction.[57] Angelina joins her friend Maria, who is 'sitting with a group of professional drinkers that were always good for the craic.'[58] Suddenly, they are interrupted by a group of sightseers who are making the rounds at 'traditional' Irish pubs:

> A large group of tourists armed with cameras came in and looked up at the cracked ceilings and black-lacquered beams. They were hushed and wide-eyed as though they'd just entered a chapel. They started clicking away at the bar and, in particular, at an elderly man in an outsize tweed jacket sucking on a pint of Guinness.
>
> 'Most photographed man in Belfast,' someone in Angelina's group commented.
>
> There was a wistful silence, a sudden unspoken communion as everyone thought of the past, sensing that all things local and Irish had been driven underground to the last few pubs, as if it was quaint history, finished with the bloodshed, swept away entirely.
>
> The tour group filed out without even taking a drink.
>
> 'Would you look at that,' muttered one of the men at her table, offended, and he snapped them with his mobile phone in mockery as they left.[59]

Here Angelina partakes in a different kind of fantasy—that of commodity fetishism. In this scene, the tour group leaves 'without even taking a drink' because the intoxicant is the phantasmagoria of 'post-conflict' Belfast that the city projects to its visitors. As Pauline Hadaway points out, after the Agreement 'political objectives remain constant: to reinvent postconflict Belfast as normal, if not placeless, and to make it economically viable.'[60] The city, eager to be 'finished with the bloodshed' of the Troubles, normalises its history as '"heritage" (a Peace Process euphemism)' in a process of political whitewashing.[61] With this transfigurative gesture, Belfast's history is 'bundled safely into a kind of distanced quaintness attuned to its function as a spectacle of and for tourism.'[62] Through the prosthesis of the camera, the tourists view Kelly's Cellars and its patrons as a spectacle—albeit, a different form of spectacle than the Canadian tourist's experience of scopic terror while recording the riots in 'Taking Care of Business.' Rather, the tourists who visit Kelly's Cellars record a moment of inaction whereby the patrons become the static objects of an exhibition. As John Berger argues, '[t]he spectacle creates an eternal present of immediate expectation: memory ceases to be necessary or desirable. With the loss of memory the continuities of meaning and judgement are also lost to us. The camera relieves us of the burden of memory … for the camera records in order to forget.'[63] Thus, the tour group experiences Belfast at an aesthetic remove and is unable to engage closely with the city or the memory of its fraught history.

Jenkinson disagrees with the use of the term 'post-conflict' to describe contemporary Belfast, stating, 'It just isn't used in normal parlance. It serves its function in certain discourses as a form of branding.'[64] Similarly to the Guinness injunction to 'BELIEVE IN YOURSELF,' this rebranding of the city as 'post-conflict' is a process of 'image-building.'[65] Belfast is building a new self-image as part of its larger redevelopment project. Angelina also seeks a new self-image; therefore she allows herself to participate in the spectacle at Kelly's Cellars as it is framed through the tourists' viewfinders. In the new tourist culture of Belfast, the tourist-photographer is a scopically driven consumer. The surveillance culture that pervaded the militarised space of Belfast during the Troubles has evolved in the post-Agreement period, becoming commercialised through Belfast City Council's 'Troubles tourism' infrastructure. By documenting the city through their photographic gaze, tourists become complicit with the Agreement. There is a structural homology between the text of the Agreement and tourists' photo-documentation of contemporary Belfast. Tourists therefore become counter-signatories of the Agreement, and their photographs attest to its effectiveness in creating a 'post-conflict' society.

'After the Peace'

This process of retroactive image-building also occurs in Jenkinson's story 'After the Peace,' which is set in 2013, several years after 'Scenes from an Empty Attic' takes place. It tells the tale of Jim Porter, a former IRA bomber who was imprisoned from 1985 to 1998 for bombing a London hotel. As Graham observes, 'prisoner release, decommissioning, and policing have all been seen as issues which disrupt the [Peace] Process; yet because they were all issues within the Process, they were already encapsulated by the Process at its outset.'[66] Jim has been out of prison for fifteen years, but he remains obsessed with the period of his incarceration and he is unable to move on. Jim still hangs about Madden's Bar, a local republican haunt, along with the 'old boys,' other former IRA men.[67] There he takes meetings with journalists and dramatists and sells them his story. These dealings render him an outcast among his former IRA comrades, who resent him for being a sell-out. He reflects, 'People liked mystery. They only respected the shadowy gunmen who'd never been convicted. Maybe after his death they'd canonise him, but for now they hated him for selling his life story (a kill 'n tell, they called it blackly). He had to be careful where he went in the city.'[68] Nonetheless, Jim tells himself that his actions are justified because they are commercially driven: 'The only reason he was talking to these arty headcases was for the pennies.'[69] He meets with Jack Hyams, a posh London playwright and director who pays Jim to serve as a consultant for his play about the Troubles. When Jack enters the bar and begins to speak to him in 'an Oxbridge accent,' Jim is immediately 'aware that everyone in the bar had turned round, their eyes trained on the man like so many gunsights. They rarely clocked foreigners … but their ears honed in like radars on English accents. Old habits died hard.'[70] Jim takes advantage of the residual surveillance culture in Belfast, as it provides spectators for his nostalgic performance of paramilitarism.

Jack and his assistant Melanie buy Jim a continuous supply of pints to loosen his tongue until he is 'lamming them down.'[71] When Jim makes seemingly offhand remarks about his experience, Jack is eager for more and he prompts Jim with questions such as, 'Can we just unpack that a bit?'[72] He tries to pry as much information as possible out of Jim by adopting a conciliatory tone:

> 'The most important thing,' said Jack, 'is that you're comfortable with this process. We're looking for the authentic voice in this project, so please, talk as naturally and as freely as you can.'

> 'This project [...] this process,' repeated Jim to himself. It was only a poxy play in some obscure London studio that the liberal elite would fawn over and give great reviews—'its searing unpeeling of the truth', or some such bollocks. He nodded politely although he'd heard all this a lump of times.[73]

From Jim's perspective, these potted descriptions of the interview as a 'project' and a 'process' echo the platitudes that were used to explain the peace process. Having given similar interviews before, Jim comprehends that this is a performance in which he will serve as the 'authentic voice' of the IRA bomber. Later, his performance in the pub will be recapitulated onstage before a bourgeois audience of London theatregoers. In the post-Agreement period, paramilitarism becomes a commodity culture for the London metropolitan consumer. Jim discloses that his identity at the time when he planted the bomb was also a self-conscious performance: '"I've a room booked under Gerry Adams," he'd said. No one in London knew that Gerry Adams was commander-in-chief of the IRA and it would be a great joke to tell the fellas when he got back to Belfast.'[74] He continues, 'In jail he hadn't mixed. He was different to them with all their Jailtacht bollocks, and he didn't care about the mythopoeic world of the famine and the six hundred years of hurt. He only cared about modern injustices.'[75] Jim attempts to emphasise his singularity and apartness from the other IRA inmates at the Long Kesh/Maze Prison. However, his mimicry of Gerry Adams when checking in at the London hotel suggests that he and his fellow bombers merely served as proxies for the IRA commander-in-chief who gave them their orders—stand-ins for the star actor.

Once the interview wraps up, Jack is keen to pose some further questions off the record. He asks, '"What do you think of the peace?" "I don't give a fiddler's," said Jim with shining eyes. He was bladdered and he was ready to deliver his own killer blow. "You see, I'm from London."'[76] Jack and Melanie are visibly disappointed by this revelation and they depart immediately, with Jack calling loudly, 'The cheque's in the post.'[77] This unsettles Jim momentarily, but once they are gone he revels in their displeasure at his unexpected provenance. He thinks, 'Their faces when he said he was English! His mother had been Ballymurphy-born, right enough, but his father had been English ... Just a wee quirk of fate that he ended up in Belfast at all.'[78] He continues, 'But it was one over on the English, wasn't it? The irony—that the English kept longing to hear how you killed the English—wasn't lost on him.'[79] However, Jim's shifting self-identification between (Northern) Irish and

English, depending on the circumstances of his story, reveals his liminal sense of identity. This performance is not intended solely for Jack and Melanie—the entire time he is speaking to them, Jim is aware that he is also being watched by a table of American tourists. He comments, 'A couple of Yankee tourists were sitting opposite with their half Guinnesses apiece. Nobody ever paid the tourists a bit of mind. Americans thought all they needed to be Irish was an O at the beginning of their surname. Sure the Dubliners gave Michelle Obama a certificate to say she was Irish. O'Bama, eh?'[80] Jim pretends to ignore the Americans, but he is cognisant that they do not come to Madden's Bar so much for the taste of a Guinness half pint as for a taste of 'traditional' republican pub culture. After Jim's conversation with Jack and Melanie concludes, Jim observes, '[t]he two fat Americans disengaged themselves from their table and waddled towards the door. They didn't know they had to press the buzzer to get out. Why, even the stupid buzzer was just another relic from the Troubles. But paranoid men like Sean had campaigned for it to stay.'[81] Jim's dismissal of the buzzer and of former IRA operative Sean as 'relics' from the Troubles is ironic considering that he also feels like a cultural artefact.

In addition to his English interviewers and the American tourists, Jim is also being watched by his former IRA comrades. Sean approaches Jim with a look of malice, and 'the bar collectively [holds] its breath.'[82] 'You better not have told the English about the Glenmachan Road,' he warns.[83] Jim replies, 'I never did. No.'[84] The onlookers in the pub 'all eagerly lit up, relieved that it was only the wind that had kicked off that night. Jim sat where he was, hardly aware of them, smothered in the company of ghosts. This last pint tasted a bit bitter, but then he'd had so many his stomach was surely turned. The smoke reminded him of the old days.'[85] Jim indulges in paramilitary nostalgia for the ostensible benefit of his mixed audience in the pub, but it is apparent that he mainly does this for the sake of having something to do. He recalls, 'An English journalist had once written about the blocks and jabs of Jim's hands, "reflecting the impasse of his wasted life." He fell silent a second.'[86] After his prison release, Jim does not know what to do other than relive his story—he cannot return to his former life as a bomber due to decommissioning. He is also unable to assimilate into contemporary society, which indicates the failure of state reintegration schemes for former paramilitary prisoners. Although this tale is set 'After the Peace,' Jim cannot find solace because he is at an existential impasse. After the Agreement, his place in the new Belfast is uncertain and he is unsure of what to do with his life. So he spends it at the pub surrounded by his former comrades, for 'Madden's [is] his home now.'[87] Jim does not feel at home in himself due to his confused sense of identity; thus, he prefers to be 'smothered in the company of ghosts.' His predicament demonstrates the general need for catharsis within a putatively 'post-conflict' society.

Conclusion

Contemporary Belfast projects its 'post-conflict' self-image onto the screen of a polished, invented cityscape which exists in the arena of global commerce. 'Visit Belfast,' the City Council's official tourism website, pronounces that 'Belfast lays claim to a unique history. And there's something for everyone to love. Shopping, tours, world-class dining, block rockin' beats, you name it—we've got it. So dig a little deeper and start planning your Belfast adventure!'[88] On their 'About Belfast' page they ask, 'So what's Belfast like today? Well, it's packed with history.'[89] Indeed, the city's history is 'packed' into its museums, tourist gift shops, and pubs as 'archival curio, cultural treasure, or commodified and reified remnant.'[90] As Jenkinson indicates in her short fiction, this has a homologous effect upon the ways in which tourists perceive Belfast's inhabitants—often as visual artefacts that they consume through the lens of the camera. In this scenario, as Aaron Kelly asserts, history is 'a pre-packaged commodity designed to prophesise retroactively a present which, although promising diversity, delivers nothing but the selfsame capitalism without alternatives.'[91] The commercialisation of Belfast's history reflects the clean break with the past which was proposed by the peace process and subsequently brokered by the signatories of the Good Friday/Belfast Agreement. The Downing Street Declaration of 1993 expresses the hope that the peace process will 'remove the conflict' and 'overcome the legacy of history.'[92] Five years later in 1998, the Agreement affirms that 'we can best honour' the 'regrettable legacy of suffering' in Northern Ireland 'through a fresh start.'[93] When Jenkinson 'digs a little deeper' in her stories, it becomes apparent that this shiny version of Belfast displaces the terrain of struggle for the local residents whom it alienates. She probes beneath the city's smooth facade in order to study the textural variability of lived cultural identities. These are inflected by Belfast's history of conflict, which the city sublimates and 'lays claim to' only insofar as it is 'unique.'

The city's new identity reflects this self-sustaining, ahistorical, and positivist metanarrative, which overcodes the individual narratives of its denizens. As Graham argues, the peace process 'has always found itself coming back to the undefined issue of identity.'[94] The proleptic discourse of the Agreement glosses over the complexities of identity as a lived process by calcifying it into 'two communities' which are framed in mutual opposition via the consociational model of government. Jenkinson indicates in her short fiction that as a consequence, 'the real business of this city was never finished.'[95] Correspondingly, Hadaway maintains that the 'prevailing mood of the "post agreement" years became one of expectancy: watching and waiting for new social and cultural arrangements to emerge from the maze of unresolved antagonisms that informed the Northern Ireland Peace Process. Through years of constructive dialogue, punctuated by crisis and dissolution, the awareness of being "in process" somehow leant a sense of forward movement to a period of suspension.'[96]

This politics of non-resolution registers homologically in post-Agreement literature, and Jenkinson's writing exhibits this effect. The individual trajectories of her characters demonstrate a forestalment that parallels the contemporaneous phase of political suspension. This is due to the fact that, as Miriam Gamble notes, there is 'a debilitating lack of framework; the absence of a clear backdrop against which to "define" [oneself].'[97] This sense of liminality continues to shape the present moment since, as Birte Heidemann points out, 'an end to the post-Agreement period has yet to be defined.'[98] Similarly, Jenkinson observes, 'You know, this could be another "pre-conflict" era, if it happens again. So maybe we're actually in a "mid-conflict" lull. How long does "post-conflict" go on? It could be another hundred years.'[99] In her 2004 debut collection, she explores the *Contemporary Problems* of the emergent post-Agreement culture. Over a decade later, her tales in *Aphrodite's Kiss* suggest that Northern Irish society is still unable to find its moorings. This is compounded by the fact that Belfast's self-identity is totally outward-facing, directed towards its visitors and investors and away from its citizens. Hence the price of peace is to become a seemingly neutral, depoliticised space of consumerism whose governing logic is that of global capital. Jenkinson intervenes in this attempted erasure of place via an urban aesthetic of dissent that recontextualises the city as a contested site. Her Belfast stories demonstrate an ironic realist mode which is conscious of the city's image-in-circulation but which is distanced from it. These texts register Belfast's dislocationality while juxtaposing it with familiar, localised imagery and language in order to reimagine the post-Agreement city as self-divided but also self-aware.

Notes

1. Rosemary Jenkinson, 'Banter and booze – you can't write about Belfast without them,' *The Irish Times* (24 June 2016), n.p.
2. Aaron Kelly, 'Walled Communities,' in Eoghan McTigue, *All Over Again* (Belfast: Belfast Exposed Photography, 2004), n.p.; Rosemary Jenkinson, *Contemporary Problems Nos. 53 & 54* (Belfast: Lagan Press, 2004), p. 68.
3. Ibid., p. 369.
4. See Eamonn Hughes, 'Introduction: Northern Ireland – Border Country,' in Eamonn Hughes, ed., *Culture and Politics in Northern Ireland, 1960–1990* (Milton Keynes: Open University Press, 1991), pp. 1–12.
5. Colin Graham, '"Let's Get Killed": Culture and Peace in Northern Ireland,' in Wanda Balzano et al., eds, *Irish Postmodernisms and Popular Culture* (Basingstoke: Palgrave Macmillan, 2007), pp. 171–183; p. 174.
6. Jenkinson, 2016, n.p.
7. Graham, 2007, p. 176.
8. Jenkinson, 2016, n.p.
9. Jenkinson, 2004, p. 85.
10. Sandy Row is a predominantly Protestant, working-class neighbourhood in Belfast. It is a staunchly loyalist area with connections to the Orange Order and the Ulster Volunteer Force (UVF). Sandy Row and Donegall Pass are very old sectarian 'communities' which pre-date the partitioning of Ireland. When the UVF in its first form was established in the early twentieth century, it had a good recruiting ground in those areas. The Ulster Defence Association or UDA (formed during the Troubles) took some ground from the UVF in parts of the city, but the former is much more prominent in newer housing estates and on the Shankill.
11. Jenkinson, 2004, p. 83.
12. Colin Graham, 'Gagarin's Point of View: Memory and Space in Recent Northern Irish Art,' *The Irish Review*, 40/41 (Winter 2009), pp. 104–113; p. 107.
13. Jenkinson, 2004, p. 84.
14. Jenkinson, 2004, p. 88.
15. Dawn Miranda Sherratt-Bado and Rosemary Jenkinson, 'Rosemary Jenkinson Interview: "Belfast dialect is like Synge on acid,"' *The Irish Times*, 3 October 2016, n.p.
16. PSNI stands for the Police Service of Northern Ireland. Operation Banner, the British Army's operation in Northern Ireland, was the longest continuous deployment in the history of the British military. The operation was initiated in 1969 and it continued until July 2007.
17. Jenkinson, 2004, p. 85.
18. Ibid.

19. Ibid., p. 86.
20. Ibid., p. 87.
21. Ibid., pp. 88–89.
22. Ibid., p. 89.
23. Ibid., pp. 91–92.
24. Ibid., p. 93.
25. Ibid., p. 93.
26. Ibid., p. 88.
27. Colin Graham, '"Every Passer-by a Culprit?": Archive Fever, Photography and the Peace in Belfast,' *Third Text*, 19.5 (2006), pp. 567–580; p. 572.
28. Personal correspondence, 28 September 2016.
29. Jenkinson, 2004, p. 69. Taig is an offensive term (chiefly in Northern Ireland) used in reference to a Catholic or Irish nationalist.
30. Ibid., p. 67.
31. Ibid.
32. Ibid., p. 68.
33. Liam Kelly, *Thinking Long: Contemporary Art in the North of Ireland* (Oysterhaven: Gandon Editions, 1996), p. 59.
34. Jenkinson, 2004, p. 69.
35. A. Kelly, 2004, n.p.
36. L. Kelly, 1996, p. 8.
37. Jenkinson, 2004, p. 69.
38. Ibid., p. 71.
39. Liam O'Dowd, 'Belfast Transitions,' in Pauline Hadaway, ed., *Where are the People? Contemporary Photographs of Belfast 2002–2010* (Belfast: Belfast Exposed Photography, 2010), pp. 22–37; p. 31.
40. Jenkinson, 2004, 69.
41. Daniel Jewesbury, 'Nothing Left', in Pauline Hadaway, ed., *Where are the People? Contemporary Photographs of Belfast 2002–2010* (Belfast: Belfast Exposed Photography, 2010), pp. 38–45; p. 39.
42. Jenkinson, 2004, p. 71.
43. Graham, '"Every Passer-by a Culprit,"' p. 572.
44. http://www.diageo.com/en-us/ourbusiness/aboutus/Pages/default.aspx.
45. Ibid.
46. Colin Graham, '... maybe that's just Blarney': Irish Culture and the Persistence of Authenticity,' in Colin Graham and Richard Kirkland, eds, *Ireland and Cultural Theory: The Mechanics of Authenticity* (Basingstoke: Macmillan Press Ltd., 1999), pp. 7–28; p. 25. This essay features a relevant discussion of a television advertisement for Smithwick's ale, which is also produced by Diageo.

47. Dawn Miranda Sherratt-Bado, 'Storied Women,' *Dublin Review of Books*, 86 (February 2017), n.p.
48. Jenkinson, 2004, p. 69, 71.
49. Aaron Kelly, 'Geopolitical Eclipse,' *Third Text*, 19.5 (2005), pp. 545–553; p. 548.
50. Ibid., p. 547.
51. Ibid., p. 549.
52. Jenkinson, 2004, p. 70.
53. Richard Kirkland, *Literature and Culture in Northern Ireland Since 1965: Moments of Danger* (London: Longman, 1996), p. 7.
54. Ibid., p. 9.
55. Jenkinson, 2015, p. 12.
56. Ibid.
57. Jenkinson, 2016, n.p.
58. Jenkinson, 2015, p. 10.
59. Jenkinson, ibid., pp. 10–11.
60. Pauline Hadaway, ed., *Where are the People? Contemporary Photographs of Belfast 2002–2010* (Belfast: Belfast Exposed Photography, 2010), p. 7.
61. Jenkinson, 2015, p. 11; Graham, 2009, p. 105.
62. Graham, ibid.
63. John Berger, 'Uses of Photography,' in Geoff Dyer, ed., *Understanding a Photograph* (London: Penguin, 2013), pp. 49–60; p. 55.
64. Sherratt-Bado and Jenkinson, 2016, n.p.
65. Nicholas Allen and Aaron Kelly, eds., Introduction, *The Cities of Belfast* (Dublin: Four Courts Press, 2003), pp. 7–18; p. 16.
66. Graham, 2007, p. 177.
67. Jenkinson, 2016, n.p.
68. Jenkinson, 2015, p. 127.
69. Ibid., p. 123.
70. Ibid.
71. Ibid., p. 127.
72. Ibid., p. 130.
73. Ibid., pp. 124–125.
74. Ibid., p. 126. Claims that Gerry Adams, former President of Sinn Féin, was an IRA commander-in-chief have not been proven.
75. Ibid.
76. Ibid., p. 135.
77. Ibid., p. 130.
78. Ibid., pp. 131–132.
79. Ibid., p. 128.
80. Ibid., p. 122.

81. Ibid., p. 131.
82. Ibid., p. 132.
83. Ibid.
84. Ibid.
85. Ibid., pp. 132–133.
86. Ibid., p. 127.
87. Ibid., p. 128.
88. www.visitbelfast.com.
89. Ibid.
90. A. Kelly, 2005, p. 550.
91. Ibid.
92. *Joint Declaration on Peace: The Downing Street Declaration* (15 December 1993, para. 1).
93. *The Agreement* 1998, 'Declaration of Support', para. 2.
94. Ibid., p. 172.
95. Jenkinson, 2004, p. 70.
96. Hadaway, 2010, p. 8.
97. Gamble, '"The gentle art of re-perceiving,"' p. 365.
98. Birte Heidemann, *Post-Agreement Northern Irish Literature: Lost in a Liminal Space?* (Basingstoke: Palgrave Macmillan, 2016), p. 46.
99. Sherratt-Bado and Jenkinson, 2016, n.p.

References

Allen, Nicholas, and Aaron Kelly, eds. 'Introduction.' In *The Cities of Belfast*, 7–18. Dublin: Four Courts Press, 2003.

Berger, John. 'Uses of Photography.' In *Understanding a Photograph*, ed. Geoff Dyer, 49–60. London: Penguin, 2013.

Diageo. http://www.diageo.com/en-us/ourbusiness/aboutus/Pages/default.aspx. Accessed January 19, 2017.

Gamble, Miriam. '"The Gentle Art of Re-Perceiving": Post-Ceasefire Identity in the Poetry of Alan Gillis.' *Irish Studies Review*, 17, no. 3 (2009): 361–376.

Graham, Colin. '"… maybe that's just Blarney": Irish Culture and the Persistence of Authenticity.' In *Ireland and Cultural Theory: The Mechanics of Authenticity*, ed. Colin Graham and Richard Kirkland, 7–28. Basingstoke: Macmillan Press Ltd, 1999.

———. '"Every Passer-by a Culprit?": Archive Fever, Photography and the Peace in Belfast.' *Third Text*, 19, no. 5 (2006): 567–580.

———. '"Let's Get Killed": Culture and Peace in Northern Ireland.' In *Irish Postmodernisms and Popular Culture*, ed. Wanda Balzano et al., 171–183. Basingstoke: Palgrave Macmillan, 2007.

———. 'Gagarin's Point of View: Memory and Space in Recent Northern Irish Art.' *The Irish Review*, 40, no. 41 (Winter, 2009): 104–113.
Hadaway, Pauline. 'Introduction.' In *Where Are the People? Contemporary Photographs of Belfast 2002–2010*, ed. Pauline Hadaway, 7–9. Belfast: Belfast Exposed Photography, 2010.
Heidemann, Birte. *Post-Agreement Northern Irish Literature: Lost in a Liminal Space?* Basingstoke: Palgrave Macmillan, 2016.
Hughes, Eamonn. 'Introduction: Northern Ireland—Border Country.' In *Culture and Politics in Northern Ireland, 1960–1990*, ed. Eamonn Hughes, 1–12. Milton Keynes: Open University Press, 1991.
Jenkinson, Rosemary. *Contemporary Problems Nos. 53 & 54.* Belfast: Lagan Press, 2004.
———. *Aphrodite's Kiss and Further Stories.* Belfast: Whittrick Press, 2015.
———. 'Banter and Booze—You Can't Write About Belfast Without Them.' *The Irish Times*, June 24, 2016. http://www.irishtimes.com/culture/books/rosemary-jenkinson-banter-and-booze-you-can-t-write-about-belfast-without-them-1.2698502. Accessed June 24, 2016.
Jewesbury, Daniel. 'Nothing Left.' In *Where Are the People? Contemporary Photographs of Belfast 2002–2010*, ed. Pauline Hadaway, 38–45. Belfast: Belfast Exposed Photography, 2010.
Joint Declaration on Peace: The Downing Street Declaration (December 15, 1993), para. 1. http://cain.ulst.ac.uk/events/peace/docs/dsd151293.htm. Accessed January 24, 2017.
Kelly, Liam. *Thinking Long: Contemporary Art in the North of Ireland.* Oysterhaven: Gandon Editions, 1996.
Kelly, Aaron. 'Walled Communities.' In *All Over Again*, by Eoghan McTigue. Belfast: Belfast Exposed Photography, 2004.
———. 'Geopolitical Eclipse.' *Third Text*, 19, no. 5 (2005): 545–553.
———. 'Introduction: The Troubles with the Peace Process: Contemporary Northern Irish Culture.' *The Irish Review*, 40, no. 41 (Winter, 2009): 1–17.
Kirkland, Richard. *Literature and Culture in Northern Ireland Since 1965: Moments of Danger.* London: Longman, 1996.
O'Dowd, Liam. 'Belfast Transitions.' In *Where Are the People?: Contemporary Photographs of Belfast 2002–2010*, ed. Pauline, Hadaway, 22–37. Belfast: Belfast Exposed Photography, 2010.
Sherratt-Bado, Dawn Miranda. 'Storied Women.' *Dublin Review of Books*, 86 (February 2017). http://www.drb.ie/essays/storied-women. Accessed February 1, 2017.
Sherratt-Bado, Dawn Miranda, and Rosemary Jenkinson. 'Rosemary Jenkinson Interview: "Belfast Dialect Is like Synge on Acid".' *The Irish Times*, October 3, 2016. http://www.irishtimes.com/culture/books/rosemary-jenkinson-interview-belfast-dialect-is-like-synge-on-acid-1.2814719. Accessed October 3, 2016.

The Agreement: Agreement Reached in the Multi-Party Negotiations (Good Friday Agreement), issued 10 April 1998, 'Declaration of Support,' Paragraph 2. http://cain.ulst.ac.uk/events/peace/docs/agreement.htm. Accessed January 24, 2017.

Visit Belfast. www.visitbelfast.com/. Accessed January 24, 2017.

CHAPTER 12

The City of the Farset: Portrayals of Belfast in Three Novels by Glenn Patterson

Terry Phillips

Recent writing on the concept of the city has focused, among other ideas, on the urban landscape as a place of change. Such change has several manifestations, including urban expansion outwards into hitherto rural or semi-rural neighbourhoods, the change in the social make-up of particular parts of a city and the introduction of new developments such as major changes to the transport infrastructure. Awareness of the city as a changing entity brings with it an awareness of the varied interactions between cities and the individuals who live in them, including both those individuals' perceptions of the city they inhabit and their relations with one another. Less attention has been given to what might be termed 'sub-regions' of the city, yet it is within these that the interaction of individuals within a community becomes most significant.[1] This chapter addresses, against the background of constant change, the nature of the communities within individual regions of a city and the exercise and nature of power both within and beyond such communities.

The novels of Glenn Patterson, set in his native city of Belfast, provide insights into urban life, both generally, as it is lived in any modern Western city, and also in this particular industrial city in the beautiful surroundings of northeastern Ireland.[2] The three novels on which this chapter focuses

T. Phillips (✉)
Department of English, Liverpool Hope University, Liverpool, UK

M. Beville, D. Flynn (eds.), *Irish Urban Fictions*, Literary Urban Studies, https://doi.org/10.1007/978-3-319-98322-6_12

each illustrate some aspect of urban living as a place of change and shifting relationships, including relationships of power. Two of them, *Burning Your Own*[3] and *Number 5*,[4] focus on life in outlying estates and explore the networks of relationships to be found there. *Number 5* has a particular focus on the changing landscape and increasing urbanisation of outlying districts. Power relationships, always a feature of urban life, figure in varying degrees in all three novels but are particularly important in *The International*,[5] the only one of the three with a city centre location, a location which, to a greater degree than that of the other two novels, awakens an imaginative response in its readers. Patterson has made it clear that he sees the 'Troubles' as only a part of the story of Belfast.[6] *Burning Your Own* and, to a lesser extent, *The International* can be read as Troubles fiction, and there have been some insightful analyses of this aspect.[7] The following analysis, while not ignoring the impact of the Troubles, focuses on aspects of the urban, both in relation to Belfast and a more general western European context.

Burning Your Own: Power Struggles on the Larkview Estate

Burning Your Own, Patterson's first novel, published in 1988, is mainly set in the Larkview Estate on the outskirts of the city. The estate, although fictional in name, is, as Patterson has made clear,[8] intended to represent a phenomenon emerging before and during his own childhood, not only in Belfast but in other cities: the new housing estates, built on the edges of cities, and representing one aspect of that shift outwards in urban landscapes. Such shifts often, as in this case, enlarge the area of the city, as well as bringing its boundaries closer to outlying towns such as the fictional Derrybeg. Larkview may be regarded as representative of such estates. The shopping precinct, typical of unimaginative 1960s architecture, has only five shops, 'each a division of a single, square, one-storey building, each with the same blank, glassy front, surrounded by breeze-blocks and smooth, purplish brick'.[9]

While such estates, sometimes described as 'overspill' estates are typical of shifts in the urban landscape in the second half of the twentieth century, the phenomenon of the suburb has a longer history. In the second part of the novel, the ten-year-old Mal, the central character of the story, and his mother stay for some time in an older suburb, the home of her brother Simon and his family, as a consequence of his father's drinking. The location of this house is on the city's north side, a mile and a half from the Shankill.[10] Mal's father had succeeded in moving into this area, but the collapse of his business had forced him to leave for Larkview. His fate illustrates a particular aspect of sub-regional belonging, in this case movement out of and into such sub-regions, often an indicator of social mobility.

In the third part of the novel, after Mal's return to Larkview, his father begins to find some respectability in the community, which comes from his securing of employment, working for his brother-in-law Simon, who has obtained a Housing Trust contract to build on the site of the woods close to the estate. It is however respectability within an ultimately powerless community. The further expansion of the ever-changing urban landscape will be generated by those who have power in the city and who themselves live outside Larkview, people like Simon. Such people are one element in what some commentators on city life describe as the urban regime, the 'informal partnership' of political and economic elites.[11]

The central action of *Burning Your Own* is set in July and August 1969, coinciding with the sending of the British army into Belfast, and is recounted largely from Mal's perspective. His relationship to the local gang is very much that of an outsider, mirroring his father's status in relation to the Larkview community and Larkview's status in relation to the city of Belfast. Mal forms a friendship with Francy Hagan, a Catholic, who is even more of an outsider than Mal and who spends most of his time on the dump on the edge of the estate. This is the main focus of readings which treat the novel as Troubles fiction. However the gang relationships may be seen as metonymic, as a microcosm of the adult community of Larkview, with the younger generation reflecting something of the power relationships of the adults based on economics and social standing. In the first part of the novel, 15-year-old Mucker dominates the estate, creating sometimes irrational loyalty, which reflects some aspects of political dominance in the adult world. Nevertheless, there are limits to the acceptance of irrationality, and Mucker's behaviour on Eleventh Night, after they discover that the centrepole they have hidden has been burnt, results in Andy Hardy taking over the leadership of the gang.[12]

Francy Hagan is a key figure, introduced in the first line of the novel, before any of the other characters, including Mal: 'In the beginning'—said Francy—'was the dump',[13] making a profound claim for the dump, an area in itself liminal, since it borders the woods which separate the estate from Derrybeg. Mal finds acceptance difficult, possibly reflecting his father's low status as an out-of-work alcoholic, and he is drawn to Francy as a fellow in exclusion, having witnessed his outsider status in an early confrontation with the local refuse collectors over the fate of a rat, another excluded entity. However, in Part Three, having returned from his uncle's house, he does find a degree of acceptance. He has had his hair cut fashionably and indulged in smoking, and, significantly, his father has a job—thus once more the power relations of the children and teenagers reflect those of the adults.

The gang culture of the streets of Larkview can thus be seen as a representation of the relationships and struggles within the adult world, but in one respect it is more than representational. This is the building of the new houses on the site of the woods, for which Mal's Uncle Simon has obtained the contract. The reader learns that the often-postponed construction is about to start, during the account of the discovery of the burnt centrepole on Eleventh Night, when Andy takes over the gang leadership from Mucker. The description includes a reference to two workmen erecting a new Housing Trust sign, 'Commencing 26 July 1969'.[14] The line immediately following 'Bastards! Bastards! Bastards!' records Mucker's response to the burning of the centrepole but it could be applied to the notice, ultimately of much more significance. The woods and the adjoining dump may be seen as places with special meanings for the young people whose playground (in the broadest sense of that word) they are. It is not too fanciful to apply the idea about urban living that 'Places are imbued with affective connotations, charged with emotional and mythical meanings; the localised stories, images and memories associated with place provide meaningful cultural and historical bearings for urban individuals and communities'[15] to this, as yet undeveloped, land which has acquired its own very particular meaning for these individuals. It has significance too for their parents, altering as it will the landscape within which they live, which will in turn have its own special meaning for them. Thus those who live outside the community, those who wield power from the centre, have a significant influence on the lives lived in this estate inhabited by the powerless.

Towards the end of the third chapter of Part Three, the final upheavals begin with the return of Mucker to dominance in the gang, to be shortly followed by the petrol bombing of the Catholic chapel and the Housing Trust site, for which Mucker is originally and falsely blamed. This climax represents a confirmation of the intrusion of sectarianism into the relationships on the estate, already indicated by exclusions from the football team and the expulsion of one of the families. Nevertheless, it is important to note that it is not just about sectarianism. The attack on the Housing Trust site suggests a response to the economic power which is transforming the landscape of Larkview.

Mal no longer feels threatened by the gang. He considers that he belongs to a world where 'Francy made the rules, and everything was no more or less than he said it was'.[16] However, the novel ends with the eviction of Francy's family from the estate, and his death on the dump, after attempting to sell his second-hand assets and, in the process knocking over the braziers, causing a fatal fire. Shortly before this, Mal has visited the dump, and realising that he has now, as it were, inherited it, he concludes, 'No, he was not yet ready; in all probability he never would be'.[17] This comment, in the closing stages of the novel, suggests not only Mal's powerlessness but that of the people among whom he lives.

The novel's focus on this particular urban community in Larkview demonstrates its exclusion. Its geographic separation from the city centre is much less significant than its separation from power. The changing relationships of the youngsters on the estate, the shifting patterns of power and the changing exclusion and acceptance of individuals reflect similar relationships among the adults of the estate and demonstrate that the estate itself is a distinctive community. However, in the wider context of the whole city, such power shifts are meaningless. These are a people without power, who have no influence on the decisions which change their surroundings.

Number 5: One House, Changing Occupants

Although the powerlessness of the community which is the subject of *Number 5* is not so extreme, it is nevertheless significant, as becomes apparent as the novel progresses. It focuses on a single house, built on an estate which was originally an extension of the city boundary but eventually, several decades later, an integral part. The house, with its sequence of inhabitants, reflecting the theme of change, is the main character in this story, which is divided into five episodes, covering the years 1958–2001, and a sixth in which Ivy Moore, an ever-present character, reflects on the passage of time.

The estate, unlike Larkview, is never actually named, although it manifests some, but not all, of the almost village-like characteristics of the other estate. Unlike the houses in Larkview, these houses are for sale and clearly reflect an attempt to satisfy a market for home ownership at comparatively low cost. Both estates illustrate an often neglected but very important aspect of urban living: the areas away from the centre of a city, where in fact most of its inhabitants live. Patterson has drawn attention to this in relation to the Shankill Road area:

> They still felt like a community apart. In fact, they felt they had more in common with people immediately on the other side of the wall than they had with people in the city centre. The new clubs and bars and new hotels that were opening—they weren't going to them—and they felt that the people on the other side of the wall weren't going to them either. So in a sense there was that common experience and both of them felt that the city centre was another world, another Belfast.[18]

This comment illustrates an essential characteristic of a community which inhabits a sub-region of a city, an ambiguity in relation to the centre. There are those who live in the district in which this house is to be found who will routinely travel to the city centre and those who, like the people referred to in Patterson's comment, will not. Likewise there will be those who enjoy its closeness and those who do not.

The city of Belfast is never referred to by name in the novel, for reasons which the author has made clear, 'when you use the name "Belfast" in a work of fiction it carries with it an idea of, "Ah, ha! It's Belfast"'.[19] There are two dimensions to his comment. One is the tradition in some parts of Ireland, conscious of its rich Celtic inheritance, of looking down on the predominantly industrial city, an attitude which some of the city's own residents have assimilated and which is manifested by the fifth narrator, Mel, who, recalling a time when he lived elsewhere, comments, 'I hadn't spent all that fucking time educating myself out of the dump just to go back because a mate asked me',[20] an attitude reminiscent of that of Drew, the central character in *Fat Lad* who returns reluctantly to manage a branch of a bookshop in his native city.[21] The other dimension to Patterson's comment, particularly in relation to the 1970s and 1980s is, of course, the emergence of the Troubles, which for some, particularly outside Ireland, seem almost to define the city. Here, as in Patterson's other novels, the Troubles are a part, though only a part of the context of the city, figuring, for example, in some of the political discussions at the neighbouring house of András Hideg.

Each section begins with an estate agent's advertisement for the three-bedroomed terraced house exhibiting changing features of the house and its location. The descriptions demonstrate little change in the house but quite significant change in the environs, foregrounding once more that theme of constant change, as the city expands outwards. The initial advertisement advertises the house as recently reconstructed and 'pleasantly situated in healthy rural surroundings',[22] and Stella Falloon, the first narrator, describes it as consisting of only three streets. The first two of the six advertisements, which can be dated to 1958 and 1963, highlight the location. 'Healthy rural surroundings' in the first becomes simply 'excellent location' in the second,[23] and the two following suggest no need to highlight the location, which by implication has become a part of the city. The fifth promises, 'the attraction of this ever-popular development will be further enhanced by the Little Lake shopping centre […] opening June 1997',[24] demonstrating the predominant commercial considerations, which have now provided buses (see below) and shops. The name of the centre suggests that it is built on the erstwhile rural area surrounding the first few streets on the estate, where the first narrator Stella described taking her baby for a walk: 'On the last bend before the farmhouse, a gate had been let in the hedge and beyond this a hundred yards lay a pond with ducks'.[25] In contrast, Ivy's concluding vision is of 'the top storeys of the houses behind and the chimneys of the houses behind them'.[26]

As in all these novels, in the village which becomes a suburb, a community of sorts emerges. It is never tightly knit, as a rural village might be, and the relationships of the adults who come and go within the streets are of a more informal kind than those of the children of Larkview and their gangs. This, a loosely knit community based only on the fact that these people happen to live in the same place, is very much a feature of modern urban living which, as James Donald has pointed out, eschews any sense of an enduring place-rooted community.[27] Ivy is the character who more than any other unites the tales, although she never lives at Number 5. What is described as her 'insistent neighbourliness'[28] sums up a character often found in small communities, who provokes some resistance in the first narrator, Stella Falloon, who prefers a more distant relationship. It is Ivy who first mentions András Hideg, whom she refers to as 'Andy Headache', a fugitive from the Russian invasion of Hungary. A more influential character than Ivy, he begins the practice of sitting in his front garden which initially shocks the somewhat reserved Stella, although, like her neighbours, she eventually adopts the practice.

While this kind of typical suburban community might be seen as less tightly knit than a village community, a positive aspect is manifested in the ability of the neighbours to respond to the Hungarian and his ways, suggesting something of an openness and ethnic mix more likely to be found in urban than rural communities. Hideg's annual celebration of 'Old Year's Night' becomes a feature of the community. However, the third narrator, the son of a Chinese family, the Tans, includes early incidents of racial harassment, although the irrepressible Ivy soon befriends his family, and as time passes, they seem to become accepted. Diversity is also represented initially by the presence and acceptance of Catholic families in a predominantly Protestant estate. John Rennie Short comments that 'A measure of heterogeneity is necessary for a place to be a real city'.[29] This becomes more important in *The International* but there are elements of variety in the population here. It suggests that a suburb is a more heterogeneous place than a rural village but that it does not quite present the mix of people present in a city centre.

While the community does not exhibit the intense power struggles of the children of the Larkview estate, certain characters, including Hideg, do seem influential. Paul Blake, who comes to live in the neighbourhood in the early 1970s, instigates a somewhat half-hearted attempt at a neighbourhood watch scheme. There are sometimes occasions of disagreement, for example parking problems, caused by the inevitable challenges in an increasingly built-up area and in terraced properties constructed before the almost universal use of the motor car. Interestingly Catriona Eliot challenges her husband's replacing of the front garden by a parking area because she will lose 'the link with the street, a reason other than getting into the car for being out the front of the house',[30] reminding the reader of when the neighbours, following Hideg's example, move into their gardens and underlining the significance of a street in an urban community, illustrating something of what Iris Young has called 'the being together of strangers [who ....] interact within spaces they all experience themselves as belonging to'.[31] However, while there will be influential characters here, as in any suburb, real power lies outside the estate and, like most urban communities it is predominantly affected by economic power held elsewhere. The lives of the residents are dependent on the decisions of others. For example, the initial residents have a long walk to the bus terminus, but the buses are subsequently rerouted when the district is expanded with the addition of more streets, in response to increased housing needs. Both the building of the houses and the extension of the bus services are dependent on decisions of those with economic and, in the case of the buses, political power.

Number 5, like *Burning Your Own*, is a powerful representation of life in an urban suburb, although in this case a more affluent one. The main consequence of such affluence, in comparison to Larkview, is that it manifests a much greater degree of heterogeneity. Nevertheless, in spite of this greater affluence, although there are some influential figures within the community, power, as in the case of Larkview, lies on the whole beyond it. The shift in narrators emphasises the shifts in population as does the description of the constantly changing surroundings as the city expands outwards.

The International: A City Centre Location

Flux and movement are present in a different way in *The International*. Here the building is a hotel with a constantly shifting population but the surrounding area is less subject to change, although elements of change in both time past and time to come are alluded to. There is a community linked together by the hotel, though of necessity much looser knit than the communities of Larkview or the estate in *Number 5*. However, of the three themes considered in this chapter, by far the most important in this novel is power.

The International is set in 1967, the day before the inaugural meeting of the Civil Rights Association on 29 January, at The International, an actual hotel which closed in 1975.[32] Again there is a strong emphasis on the perspective of the individual. The novel is narrated in the first person, by Danny, an 18-year-old who has come to work in the hotel immediately after the murder of Peter Ward, a real-life victim of the paramilitary Ulster Volunteer Force (UVF), and also includes the personal recollections of other individuals. It recounts one day in Danny's life but includes many retrospectives, including those coming from future conversations with some of the key characters and scenes in other parts of the hotel recounted by his workmates and enhanced by his own suppositions.[33]

The range of individual stories highlights the way in which the hotel with its city centre location provides a common shared space to which a variety of different people might have recourse, creating that sense of a city as a free space which has been highlighted in writing about cities, for example, in the comment from John Rennie Short, quoted above. The variety is greater and more heterogeneous than in the streets of *Number 5* and marks a significant distinction from the Larkview Estate. People of very different backgrounds such as the would-be television puppeteer,

Stanley and the disappointed Ingrid, in search of photographs of her ex-lover's wedding contribute to this sense, as does Danny's homosexuality, and the presence of the American couple, the Vances, with whom Danny has indulged in a threesome. Such a feeling of freedom produces its own exhilaration which Danny conveys enthusiastically when he comments, 'Half past eight on a Saturday evening in a busy bar. A blessed time, I used to think—before The Four Step Inn, The Mountainview Tavern, The Crescent Bar, before Loughinisland'.[34]

Nevertheless, for all the sense of freedom, the city centre, like Larkview and the district of *Number 5*, is not without its own power structures on both a large and a small scale. The International has a hierarchy of its own which Danny has to observe, deferring to the Master, Len Gray, the hotel's bar manager, and Hugh, the manager of the Blue Bar. However, the particular importance of the theme of power in this novel is related to its setting in the city centre, in any city the location of both economic and political power. Economic power, manifested throughout the novel, is for the most part more significant than mere political power and brings with it the power to change the landscape of the city. Towards the end of the novel, as Danny makes his way home from the day at work which has formed its substance, he makes a key observation:

> I crossed the street to the back of the City Hall and peered through the gates. Somewhere in there people's broken toilets and blocked drains and leaking roofs awaited official consideration. Somewhere in there lay the plans for our B.U.M. Families were sleeping tonight in houses where no houses might be next year if the councillors who traipsed in and out of here and over the road to The International decided it. If men like my Second Cousin Clive decided it.[35]

This is an emphatic statement about the effect and location of power. In *Burning Your Own,* the absence of power is there for the reader to see and it is manifested in *Number 5* by the wait for the rerouting of the bus service. However, here the issue is made explicit. The proposed Belfast Urban Motorway, the B.U.M., is an ever present in the urban context of the novel. The proposal to build this ambitious motorway is accompanied inevitably by planning blight, with promised housing repairs repeatedly postponed.

In this, as in any city, the landscape-changing power, with its concomitant effect on people's lives, is shared between those with economic power and those whose political power, theoretically at least, rests on the decisions of voters. This is manifested by a significant group of The International's customers: Clive White, Councillor Noades, a Unionist councillor, and Fitz, a businessman from the Republic. Clive, through whose agency Danny has gained employment at The International, provides a type of a certain kind of somewhat dubious wielder of economic power, whose methods of acquiring wealth, including a preparedness to cut profits when dealing with another agent of power, the Royal Ulster Constabulary, are described by Danny when he hears that his relative is to dine with Fitz and Noades.[36] Fitz has come with a proposal to build a 'super-garage', arousing the interest of Noades, presumably because it will provide jobs, and Clive who sees it as a way of making money. The relationship between politics and business is clearly evident in Clive's offer to set up a company which will provide a Northern Irish 'front' for Fitz, to minimise the political consequences for a Unionist councillor who will support and campaign for it.[37] A passage of free indirect narrative, conveying Clive's perception of Noades' own thinking and his brand of Unionist politics, concludes with the pragmatic reflection: 'It was all about money. Anyone could get up a meeting and pack the hall to the rafters, but when push came to shove money was what the voters cared about'.[38]

Such pragmatic considerations and alliances between men with very different aims and aspirations often have detrimental effects on the lives of some, at least, of the citizens of Belfast, or indeed any city. The proposal for the super-garage, about which the three are negotiating, is for 'an area zoned for housing, where few of the residents owned cars'[39] and the prolonged planning of the B.U.M. affects people's lives in a number of ways, as has been seen. A reflection of the negative effects of schemes devised for commercial and political motives can be seen in Patterson's comments on the real-life Sunflower Pub in Belfast: 'It can only be hoped that the bar, which was crowned the best in the city last year, can be spared the wrecking ball after the area was earmarked for an extensive rebuilding programme'.[40]

In spite of its setting within a single day, contrasting particularly with the decades of *Number 5*, the changing landscape of the city is alluded to at several points in the novel. Unlike *Burning Your Own*, it is set in actual streets, which the reader with some knowledge of central Belfast can readily bring to mind. It encompasses the characters' own imaginative response and memory. Patterson's interest in the growth of the city is well attested in *The Mill for Grinding Old People Young*,[41] and in the early part of *The International*, the reader is given a glimpse of a period when the city flourished, as Fitz recalls his grandfather's memories of the opening of the City Hall in 1906, 'standing at the balustrade of the great horseshoe stairwell, craning his neck to marvel at the mosaics lining the enormous dome a hundred feet above'.[42]

However, there is another sense in which landscapes can change, that being the way they are perceived by significant numbers of people. The International Hotel is undergoing such a change at the time in which the novel is set, because of the events of the previous June, in which four barmen from the hotel had been ambushed by members of the paramilitary UVF, resulting in the death of the real-life Peter Ward, replaced at the hotel by the fictional Danny. He comments somewhat wryly, 'The International barmen. It was a sort of fame, I suppose'.[43] It is an interesting example of the way places and buildings acquire a symbolic meaning, which is an ongoing process, subject to change in any urban landscape, although political events in this particular city have caused a more rapid change. In the final chapter of the novel which provides a postscript to the events of this day, 28 January 1967, a very different Belfast is about to emerge, anticipated by a comment much earlier in the novel referring to that day being before, 'it was brought shaking, quaking and laying about it with batons and stones on to the world's small screens'.[44] The developments which would have taken place in any developing urban landscape have been given a new twist, signified by the final fate of the B.U.M. proposal:

> we got a B.U.M. of sorts eventually. Contractors on the Westlink, [...] were paid a foot at a time, so many individuals and organisations wanted them dead, for so many different reasons. It is hard to ascribe civic-mindedness to people who were blasting the fuck out of their own city, but there would appear to have been something of that in the paramilitary threats.[45]

The reference to civic-mindedness evokes the protracted debates over road schemes which are very much a feature of contemporary urban life in any city, but complicated in the case of Belfast by its particular political situation in the final decades of the twentieth century.[46]

The International illustrates the exercise of power in any city, what is sometimes referred to as the urban regime, and the way in which such power affects the lives of individuals. The urban landscape described in this novel, that of the city centre, is less subject to change than that of the Larkview estate or the suburb described in *Number 5*, although the novel does record changes in the way it is seen. However, what is very clearly represented in the novel is the heterogeneity of the inhabitants of the city centre, exhibiting a diversity not to be found in the suburbs described in the two novels previously discussed.

These three novels illustrate in various ways three important aspects of urban living in many Western European cities. One key feature is change: change to the landscape, both actual and symbolic, and in the inhabitants of particular locations. Patterson has attested his interest in change, remarking of his return to Belfast in 1994, his realisation that he understood "something about Belfast I never understood before', which is this idea that it is constantly being rebuilt'.[47] This can be seen in the growth outwards of Larkview, the building and development of the unnamed suburb of *Number 5* and the way in which perceptions of the city centre change with time. A second key feature of urban living is the existence of communities within the larger city community. Each of these novels focuses on a single area: housing estate, suburb and city centre, respectively, thus illustrating the fact that cities cannot be considered in any meaningful way as an entirety, 'a synthetic totality'.[48] The heterogeneity of the communities increases as the communities themselves become less tightly bound, illustrated by the contrast between the ethnic homogeneity of Larkview and the more diverse population of the suburb described in *Number 5*. Although the city centre lacks the communal identity of a suburb, it does manifest aspects of a loose and often changing community. The third key feature is the exercise of power, both within and beyond these communities, which also varies in proportion to the closeness of the communities. In the almost village-like community of Larkview, there is an internal hierarchy operating, but the community itself is powerless, a powerlessness made explicit by the Housing Trust expansion, and indicated symbolically by Mal's final turning away from the alternative power of the dump. In *Number 5* there is a clear hierarchy manifest among the shifting population, with figures who dominate, sometimes because of their social and economic position and sometimes because of personality, Andy Hideg and Paul Blake among the more influential characters and the ever-present Ivy the least. Nevertheless here too there is an absence of real power, with public transport and developments such as the Little Lake Shopping Centre being in the control of people outside the community. An insight into those who operate real power which lies within a combination of political and economic elites is demonstrated in *The International*, where some members of the loose community who gather around the hotel meet to negotiate over the super-garage and the B.U.M.

Notes

1. For a discussion of the somewhat neglected topic of the suburb and its relationship to the city, see Herrschel, 'Cities, Suburbs and Metropolitan Areas-Governing the Regionalised City'.
2. All Patterson's novels are set in the city with the exception of *Black Night at Big Thunder Mountain* Patterson, *Black Night at Big Thunder Mountain*.
3. Patterson, *Burning Your Own*.
4. Patterson, *Number 5*.
5. Patterson, *The International*.
6. See, for example, Hicks, 'A Conversation with Glenn Patterson', 111.
7. See, for example, Kennedy-Andrews, *Fiction and the Northern Ireland Troubles since 1969: (De)constructing the North*, 102–106, Parker, *Northern Irish Literature, 1975–2006*, 127–133, and Alexander, 'Remembering to Forget: Northern Irish Fiction after the Troubles'.
8. Hicks, 'A Conversation with Glenn Patterson', 108.
9. Ibid., 40.
10. Ibid., 113.
11. Short, *The Urban Order*, 288.
12. Patterson, *Burning Your Own*, 76–77.
13. Ibid., 3.
14. Ibid., 78.
15. Balshaw and Kennedy, 'Urban Space and Representation', 6.
16. Ibid., 211–212.
17. Ibid., 232.
18. Hicks, 'A Conversation with Glenn Patterson', 7.
19. Ibid., 111.
20. Patterson, *Number 5*, 243.
21. Patterson, *Fat Lad*.
22. Patterson, *Number 5*, 3.
23. Ibid., 3 and 71.
24. Ibid., 239.
25. Ibid., 43.
26. Ibid., 308.
27. Donald, *Imagining the Modern City*, 147–171.
28. Patterson, *Number 5*, 265.
29. Short, *Urban Theory*, 9.
30. Patterson, *Number 5*, 211.
31. Young, *Justice and the Politics of Difference*, 237.
32. Patterson, *The International*, 114–115.
33. Ibid., 169.
34. Ibid., 159.

35. Ibid., 302.
36. Ibid., 123.
37. Ibid., 171–172.
38. Ibid., 236.
39. Ibid., 172.
40. A month later, the pub was saved from demolition when the plans were abandoned, Patterson, 'In Praise of Belfast'.
41. Patterson, *The Mill for Grinding Old People Young*.
42. Patterson, *The International*, 35.
43. Ibid., 79.
44. Ibid., 61.
45. Ibid., 311.
46. Westlink, after much opposition and discussion, was eventually opened in two stages in 1981 and 1983 Johnston, 'Belfast Urban Motorway & Westlink'.
47. Patterson and Mills, 'Nothing Has to Die: An Interview with Glenn Patterson', 126.
48. Balshaw and Kennedy, 'Urban Space and Representation', 1.

References

Alexander, Neal. 'Remembering to Forget: Northern Irish Fiction after the Troubles.' In *Irish Literature Since 1990*, 272–283. Manchester: Manchester University Press, 2009.

Balshaw, Maria, and Liam Kennedy, eds. 'Urban Space and Representation.' In *Urban Space and Representation*. London: Pluto Press, 2000.

Donald, James. *Imagining the Modern City*. London: Athlone Press, 1999.

Herrschel, Tassilo. 'Cities, Suburbs and Metropolitan Areas—Governing the Regionalised City.' In *Suburbanisation in Global Society*. Research in Urban Sociology 10, 107–130. London: Emerald, 2010.

Hicks, Patrick. 'A Conversation with Glenn Patterson.' *New Hibernia Review* 12, no. 2 (Summer, 2008): 106–119.

Johnston, Wesley. 'Belfast Urban Motorway & Westlink.' Northern Ireland Roads Site, 1967. http://www.wesleyjohnston.com/roads/belfasturbanmotorway.html. Accessed April 13, 2017.

Kennedy-Andrews, Elmer. *Fiction and the Northern Ireland Troubles since 1969: (De)constructing the North*. Dublin: Four Courts Press, 2003.

Parker, Michael. *Northern Irish Literature, 1975–2006*. Basingstoke: Palgrave Macmillan, 2007.

Patterson, Glenn. *Burning Your Own*. London: Minerva, 1993a.

———. *Fat Lad*. London: Minerva, 1993b.

———. *Black Night at Big Thunder Mountain*. London: Chatto and Windus, 1995.
———. *The International*. London: Transworld Publishers, 1999.
———. *Number 5*. London: Penguin, 2004.
———. *The Mill for Grinding Old People Young*. London: Faber & Faber, 2012.
———. 'In Praise of Belfast.' *The Guardian*, Sec. Travel. March 15, 2016. https://www.theguardian.com/travel/2016/mar/15/praise-of-belfast-glenn-patterson-northern-ireland.
Patterson, Glenn, and Richard Mills. 'Nothing Has to Die: An Interview with Glenn Patterson.' In *Northern Narratives*, 113–129. Writing Ulster 6. Belfast: University of Ulster, 1999.
Short, John Rennie. *The Urban Order*. Oxford: Blackwell, 1996.
———. *Urban Theory*. Basingstoke: Palgrave Macmillan, 2006.
Young, Iris Marion. *Justice and the Politics of Difference*. Princeton: Princeton University Press, 2011.

Index[1]

[1] Note: Page numbers followed by 'n' refer to notes.

M. Beville, D. Flynn (eds.), *Irish Urban Fictions*, Literary Urban Studies, https://doi.org/10.1007/978-3-319-98322-6

The manufacturer's authorised representative in the EU is Springer Nature Customer Service Centre GmbH, Europaplatz 3, 69115 Heidelberg, Germany. If you have any concerns regarding our products, please contact ProductSafety@springernature.com

Printed and bound by CPI Group (UK) Ltd, Croydon, CR0 4YY

30/09/2026

02225439-0005